RJ

NEW PERSPECTIVES ON EASTERN EUROPE AND EURASIA

The states of Eastern Europe and Eurasia are once again at the centre of global attention, particularly following Russia's 2022 full-scale invasion of Ukraine. But media coverage can only do so much in providing the necessary context to make sense of fast-moving developments. The books in this series provide original, engaging and timely perspectives on Eastern Europe and Eurasia for a general readership. Written by experts on—and from—these states, the books in the series cover an eclectic range of cutting-edge topics relating to politics, history, culture, economics and society. The series is originated by Hurst, with titles co-published or distributed in North America by Oxford University Press, New York.

Series editor: Dr Ben Noble—Associate Professor of Russian Politics at University College London and Associate Fellow at Chatham House

JOEL BURKE

Rebooting a Nation

*The Incredible Rise of Estonia,
E-Government and the Startup Revolution*

HURST & COMPANY, LONDON

First published in the United Kingdom in 2025 by
C. Hurst & Co. (Publishers) Ltd.,
New Wing, Somerset House, Strand, London, WC2R 1LA
© Joel Burke, 2025
All rights reserved.

Distributed in the United States, Canada and Latin America by
Oxford University Press, 198 Madison Avenue, New York, NY 10016,
United States of America.

The right of Joel Burke to be identified as the author of
this publication is asserted by him in accordance with the
Copyright, Designs and Patents Act, 1988.

A Cataloguing-in-Publication data record for this book
is available from the British Library.

ISBN: 9781805263012

This book is printed using paper from registered sustainable
and managed sources.

www.hurstpublishers.com

For Mom and Dad

CONTENTS

Introduction   1

PART I
REBOOTING A NATION

1. The Birth of a Unicorn   11
2. Sonic Youth   27
3. Independence Day   37
4. A Nation Reformed   47
5. Practicing E-Judo   63

Part I: Lessons Learned   73

PART II
BUILDING A DIGITAL SOCIETY

6. The Bedrock of the E-State   77
7. Becoming E-Estonia   89
8. Playing Leaptiger   107
9. The Estonian Tech Mafia   117
10. Exporting Estonia   129
11. Web War One   143
12. Software Eats Government for Breakfast   153

Part II: Lessons Learned   161

PART III
THE FUTURE OF THE E-STATE

13. Information Age   165
14. Country-as-a-Service   179

15. Ten Million Estonians	187
16. Cloud Country	197
17. The Evolution of the E-State	205
18. Mindset Over Matter	219
Part III: Lessons Learned	229
Recommended Reading	231
Selected Interviews	233
Sten Tamkivi	
Matthew Mitchell	
Marina Nitze	
Jennifer Pahlka	
Hanah Lahe	
Siim Sikkut	
Acknowledgements	259
Notes	261
Index	321

INTRODUCTION

On a cold day in early 2018, I land in Tallinn, the capital city of Estonia, a small European Union member state tucked below Finland and bordering Russia. It's a country which happens to be the epicenter of the evolution of the modern nation-state in an age of rapid technological progress (only I wasn't aware of that just yet). Going through the Lennart Meri Tallinn Airport, named for one of Estonia's most esteemed statesmen, I get my first glimpse of what the country has to offer. The airport is the epitome of efficiency and cleanliness with fast-moving lines and highly professional English-speaking staff. There is a welcoming sense of familiarity and coziness even here in the airport terminal. Each gate has a different theme: a community library, a mini gym sponsored by a local fitness chain, an oversized chess set, and games for both kids and adults.

Unlike most airports I've visited, there is no real queue for a taxi outside the terminal. This is not surprising given just how tech-literate the country is and the fact that one of the more recent Estonian unicorns (a tech company valued at more than a billion dollars) is Bolt, a serious competitor to Uber with operations in dozens of countries across Europe, Africa, and the Americas. Fortunately, I've already downloaded the Bolt app after a meeting with one of the many Estonian expats scattered about global tech communities before my trip to Tallinn. My sim card isn't working but the airport, like many public spaces in Estonia, has fast and free Wi-Fi and I quickly find a Bolt driver.

REBOOTING A NATION

I'm in Estonia for a job interview with the e-Residency program, a "startup within the government." I can't say it's a country I'd ever imagined working in, or really thought much about other than in a few conversations with more politically inclined friends in the Silicon Valley tech community who espoused the virtues of Estonia's forward-thinking government policies and startup scene. I'd visited the country once for about twenty-four hours about a decade prior on a road trip with my father. While I had a vague recollection of the Tallinn old town, much of my memory of that holiday was about being detained at the border of Estonia and Russia. After a lively discussion about the band My Chemical Romance and a large fine paid in cash to the Russian border guards, we were then walked to the border crossing joining the two nations. From there we traveled from Narva, the largest city in the east of Estonia and one heavily populated by Russian-speaking people as well as Russian émigrés (and their descendants), to the capital, Tallinn.

Walking around the beautiful old town of the capital, which was mostly preserved from the bombings during World War II and subsequent social and cultural destruction during the Soviet occupation, I was struck by the unique nature of the country. Here was a place that on one hand was surrounded by the vestiges of the past, of repeated occupations by larger regional powers, and by the ever-present threat of neighboring Russia, yet there was a tremendous newness to the country which only a few decades prior had regained its independence and had an opportunity to begin again as a sovereign and free country. The early leaders of the nation who had been molded in the hardship of the Soviet era, faced with exile, persecution, and worse, did not take lightly the opportunity and challenge of remaking Estonia into a place that would stand the test of time as an independent nation.

As I approached the e-Residency office for my interview, I was comforted by the fact that it certainly did not look like a government building. The e-Residency team was then housed apart from other government offices to maintain its independence and unique culture. It occupied part of a former industrial facility with a great stone edifice making up the lower floors which comprised offices (including e-Residency's) and shops. Placed atop the stone structure was an

INTRODUCTION

alien-looking post-modern glass cube filled mostly with luxury apartments catering to the new elite. Walking into the office, I was greeted by a sign that exclaimed "enter e-Estonia" in garish blue neon letters. The over-the-top sign and the fact that everyone wore slippers in the office created a rather relaxed atmosphere, which made the experience somewhat less nerve wracking than the average job interview. Most of the day is a blur but I remember three things distinctly: seeing that this was an incredibly unique culture; immediately understanding that Estonia's government was far ahead of any other I'd come across and that I was sure to learn a great deal that I could take back home to America someday; and a soon-to-be colleague's quip that Estonia was small enough that anyone could be the best at something here if they wanted.

I was sold. A few weeks later I accepted the job offer and showed up to the office ready to join in the development of the next iteration of the nation-state. My title at the e-Residency program was Head of Business Development, one of those generic titles that means everything and nothing, which was apt as it suited my actual responsibilities perfectly. One day my job might be to work on a partnership with a financial institution, the next to plan a delegation to a foreign nation, and the day after to speak to a group of executives at the e-estonia showroom and explain to them exactly what was happening in Estonia. Those conversations with business and government officials formed the kernel that was planted in my mind about one day writing a book about Estonia.

There have been scores of articles written about the country, but most just scratch the surface of the real story behind Estonia's successes (and failures) and the broader implications for the modern world. However, during my time working for the Estonian government, representing the nation in international fora including at the United Nations, and working with the local startup community, I was stunned at how unique the country was. It was a place which had a story that needed to be told, and one which I believed would have ramifications on the rest of the world. The impact of my time in Estonia has weighed heavily both on my career and my understanding of the role of government, the future of the nation-state, and the urgency of solving pressing problems around pressing issues like state capacity to help preserve democracy.

In 2022, I began a job as a Congressional Innovation Fellow, facilitated by TechCongress, a bipartisan nonprofit that works to bring technical expertise to Congress. As a fellow, I was staffed to a Congressional office to advise a Member of the House of Representatives on relevant tech-related matters. While the Representative I worked for was more tech-savvy than most, seeing the dysfunction of the U.S. government up close and personal made me realize just how crucial it was to tell Estonia's story and showcase how a relatively impoverished democracy with few natural resource advantages was able to build one of the most advanced digital governments in existence in only a generation. After all, if someone else can do it, then surely America could too. In 2023, I was serving as an Artificial Intelligence Fellow in the U.S. Senate, helping my boss, Senator Mike Rounds, and the institution writ large grapple with the implications that AI had brought to the fore after the release of ChatGPT. During that time, I helped to organize a series of AI Insight Forums, which brought together tech luminaries like Bill Gates, Eric Schmidt, and Elon Musk to educate Senators and Congressional staff on the impact of AI and how to consider regulating such an important technology. That experience helped solidify my belief that it was more critical than ever to tell Estonia's story as the U.S. grappled with a new technological revolution.

Thanks to rapid technological progress, the world is changing far faster than most governments realize. Remote workers can increasingly choose where to live and work, picking their ideal jurisdiction and quickly moving facilitated by scores of newly developed "digital nomad visas" with enticing benefits from governments hoping to lure in new taxpayers or just long-term visitors spending much needed foreign currency (and while globalization and remote work are currently under threat from nationalism, war, and more, in the long term competition for elite talent will only increase). Artificial Intelligence may soon enable a single individual to create a billion-dollar company on their own, creating (or further exacerbating) enormous inequality with power and resources

INTRODUCTION

accruing to fewer and fewer individuals. Even the nature of warfare is shifting. While artillery and mass destruction have made a return in the war in Ukraine, the signs of change are everywhere. With dramatically increased use of repurposed commercial drones to drop munitions into tanks, kamikaze vessels sinking ships, and commercial satellite communications units providing connectivity to the frontlines, technology is clearly reshaping warfare and the balance of power.

Many technological advances have precipitated, or at least gone hand in hand with, political change. It remains unclear whether the advances being seen in AI and beyond will benefit large, industrial-style democracies like the United States, especially if such states continue to act like they remain in an analog era and fail to adapt. AI could enable an authoritarian leader to create a low-cost dragnet of surveillance cameras, virtual communication monitoring, and autonomous drones for enforcement that would allow them to rule a nation with limited potential pushback—no robot is going to organize a coup, not yet anyway. It seems equally plausible that AI will create tremendous wealth, especially in tech hubs like Silicon Valley. And while AI may automate away many jobs, the government could effectively redistribute a portion of the newfound wealth and provide a form of universal basic income, bringing up the standard of living for many. However, the latter scenario is predicated on government working effectively. This means not only being able to pass legislation to ensure that society can adapt to technological change but to have the capacity to deliver services and public goods. Plus, it must do so in a way that remains competitive with other systems of government, whether authoritarian, neo-medieval, a charter city administered by some other more efficient government or public-private partnership, or a monarchic network state run on the blockchain.

Today, few people in the U.S.—even many of the government officials who are ostensibly beneficiaries of the current system—would argue that the status quo is working well. Many in the West are rapidly losing (or have already lost) faith in public institutions. Given the performance of government in the modern era, it's little wonder why: government agencies can't seem to create a website on time or on budget, let alone much needed munitions for a war in the heart of Europe. Seemingly obvious problems like bringing down the costs of absurdly priced healthcare or permitting reform for nuclear energy which would facilitate the creation of abundant clean energy, are never solved because of political chicanery and special interest groups.

America is facing its Macy's moment. For decades, Macy's reigned supreme in American minds as an example of commerce done well, so permeating American culture that almost everyone knew of the company. However, the company eventually faced a classic Innovator's Dilemma scenario.[1] Macy's of the early 2000s continued to invest in the same retail strategy that had worked so well for them in the past and did not take adequate note of emerging technology trends that would shake up the entire industry, namely the emergence of e-commerce and Amazon in particular. Macy's failure to invest serious resources and potentially cannibalize their own existing retail business is understandable: who wants to be the executive who makes a crazy bet on e-commerce and spends billions on an effort that may fail? Even if Macy's had invested in digital at the time, there's no guarantee they would have been successful. Operating big-box stores was in their DNA—digital wasn't. Macy's continues to be rocked by events that seem to keep pushing the market in Amazon's favor, including a global pandemic that brought people around the world inside for months on end, with many in the West with money to burn and one obvious place to do so. While Macy's has managed to make some changes by betting big on e-commerce recently, its market capitalization pales in comparison to Amazon's and the firm's future remains uncertain.

Democracy today, at least in much of the Western world, is in as fraught a situation as the one that Macy's faced just as Amazon came onto the scene. A powerful China with global ambitions, repeated financial crises, rampant inequality, conflicts and climate change leading to global instability and mass migration, plus shifting technology trends have formed a perfect storm that could sink democracy in the Western world if political leaders do not consciously make tremendous efforts to transition to nations equipped

INTRODUCTION

to survive in a digital/AI era. Unfortunately, most public institutions and the politicians and bureaucrats in charge of them stay the course and depend on inertia rather than innovating.

This book is an attempt to share lessons from a country that is both democratic and digital, one well positioned to succeed in an era defined by artificial intelligence, rapid technological and geopolitical changes, and increased uncertainty. Despite being trapped for decades under repeated occupations, Estonia was able to become a nation that is not just home to one of the highest rates of startups and unicorn tech companies per capita in the world, but one that is an undisputed e-government leader, with 99% of government services accessible online.

The book is divided into three parts. The first explores the founding of the modern nation of Estonia, attempting to understand how the country developed such a unique and entrepreneurial culture and the factors that would later help it become an e-government and tech leader. Part II peers into the early days of the development of Estonia's e-government systems along with the policies and principles that helped to shape their creation. The book culminates with a look at how the country is adapting to rapid technological change today, the potential impacts of artificial intelligence, and a new perspective on the culture and mindset that guides the nation. Keen readers are likely to notice that the book more frequently references the U.S. than other Western nations when it comes to the need for reform. This is not to imply that there are not reforms needed in Europe, but simply a reflection of my more direct knowledge of the U.S. bureaucracy.

This is not a history book or a piece of academic research, but instead a view through the eyes of someone who has been lucky enough to live and work on four continents, for governments across the world, and who saw a glimpse of a potential future—one that may help to save the West from what seems now to be an imminent decline into irrelevance if something isn't dramatically changed in how government operates. While I have a great affinity for Estonia and its people after living and working there for some time, I have worked to make this book as unbiased as possible and to present both the good and the bad, to talk about what went wrong and what went right, and where luck played a role in the country's fortunes. It is my hope that this book will serve as a useful guide or at least an interesting case study for politicians and civil servants, as well as business leaders, entrepreneurs, and civil society advocates who are looking to change the way that government functions to make it more responsive and provide real value to citizens in the digital age.

This book is the culmination of my personal knowledge from my time working at Estonia's flagship digital program and myriad other activities in the country including serving as an advisor to a high-flying local startup run by some of the most talented entrepreneurs I've met anywhere in the world, serving as a mentor at many startup hackathons, and of course from significant amounts of research and interviews conducted over the course of developing and writing the book. I owe a great debt to the many people who took the time to educate me about Estonia, e-government, state capacity, startups, and more, as well as institutions like Stanford Library which hosts a significant repository of information on Estonia's history, the e-Governance Academy which is home to many of the world's leading experts on e-government, not just in Estonia but across the world, and *Estonian World*, an English-language publication which has helped provide monolingual speakers like myself an inside look into Estonia for many years. This book would not have been possible without the generous time given by experts, academics, political figures and civil servants who made the decisions that shaped modern Estonia, and many others including President Ilves, Dr. Arvo Ott, Professor Rainer Kattel, Jennifer Pahlka, Sten Tamkivi, Dr. Matthew Mitchell, Marina Nitze, and so many more. Special thanks to Siim Sikkut, former CIO of Estonia and author of *Digital Government Excellence*, who was incredibly generous with his time and insights.

PART I

REBOOTING A NATION

1

THE BIRTH OF A UNICORN

Home to 1.3 million people spread across a landmass more than twice the size of New Jersey with one-seventh the population, Estonia is the sort of country that most world history books gloss over due to its small size. Its honorary mention will likely come in a section about countries annexed during World War II or the fall of the Soviet Union, usually as part of a larger commentary about the Berlin Wall and regional independence movements. The capital, Tallinn, is home to a picturesque medieval old town, well preserved despite numerous foreign conquests and decades of occupation. The main entrance to the old town is via a single thoroughfare with one side of the street dedicated to small shops run seemingly exclusively by Soviet-era babushkas invulnerable to the cold, selling fresh cut flowers even in the harshest of winters. The other side of the street is occupied by shops hawking tchotchkes for tourists and locally designed clothing along with what can only be described as a Finnish take on McDonald's.

Going down the cobblestone road, travelers pass through the Viru Gates, two imposing guard towers that loom over passersby as they enter medieval Tallinn. Once inside, the main streets are surrounded by scores of restaurants and pubs, most catering to tourists and replete with garish signs and excruciatingly over-the-top costumes of bright colors and medieval period garb donned by waiters beckoning travelers into their establishments. But despite the tourist traps and the occasional fast-food joint, the old town truly does feel… old. Through narrow and winding side streets, travelers stumble upon churches and historic buildings which have stood for centuries. At Christmastime holiday markets pop up—as they have for hundreds of years with their accompanying trees, a regional tradition since the Middle Ages.[1] And often, one comes across a sign of the Soviet occupation whose inflicted traumas have indelibly scarred the country, like the War of Independence Victory Column. The column adjoins the imposing Freedom Square, an open area in the old town sometimes used for public demonstrations, but which during the Soviet era was mostly the site of propaganda events and parades for Soviet holidays.

Estonia is a country with a deep history, albeit one far too few are aware of. Even basic facts about the country are not well known outside of Europe, from the language spoken (Estonian) to the fact that it is a member of both the European Union and NATO and had initially gained independence, albeit briefly, in 1918. If people have heard of the country, it is usually either because they passed through Tallinn for a day on one of the frequent cruises ferrying tourists around the region or because they have heard that this tiny country excels in one area—tech. Step outside the old town and the latter becomes much clearer. It is not unusual to see sidewalks shared between autonomous delivery robots and pedestrians. The cars and scooters whisking travelers around the city are almost universally hailed via Bolt, the Uber competitor whose CEO, Markus Villig, is one of the youngest founders of a multi-billion dollar company.[2] Even the traditional Christmas tree has been made digital through a government-run e-service which lets people find, cut down, and then pay for the tree from their phone.[3] Everywhere there are reminders of the country's forward-thinking nature when it comes to tech: near-ubiquitous free Wi-Fi, coworking spaces and cafes catering to startups, and the occasional ATM that converts cash to cryptocurrency.

Estonia's emergence as a tech hub is miraculous. In a single generation, the country became home to the most startups per capita in Europe[4] and an e-government leader, where it was ranked first

THE BIRTH OF A UNICORN

on the UN global services scoreboard for 2022 as part of the UN's E-Government Survey[5] thanks to its work digitalizing 99% of government services. It was a far cry from the newly re-independent Estonia of 1991, which was reeling after fifty years of Soviet occupation and oppression. Before the Soviets occupied the country at the outset of World War II, Estonia had a standard of living on a par with their Nordic neighbor Finland, which sits a few short hours across the Baltic Sea by ferry.[6] But by 1988, when the Estonian independence movement started in earnest, Estonians earned just 22% that of the average Finn[7] and experienced a significantly lower quality of life. There was little investment in quality physical infrastructure, although thankfully the Soviets prioritized scientific education, especially related to cybernetics and IT, and once thriving industries and farms had been decimated through an inefficient command economy and forced agricultural collectivization.

Despite the turmoil and hardship inflicted while the country suffocated under the weight of the Iron Curtain, there was a hope throughout the nation that they could achieve greatness. There had to be for people to have something to fight for in their struggle against the Soviet occupiers. Life could get better in Estonia and the country had dreams that far outstripped its size. In Estonia, a person could talk of changing the world, building a global business, or revitalizing democracy and not be laughed out of the room. Because of this, perhaps it is no surprise that Estonia has become a land of unicorns—at least a land of tech unicorns—with ten companies valued at more than $1 billion having been built in the country or by Estonian founders with more on the horizon as of late 2024.[8] The first unicorn to be born in Estonia was Skype, the global communications firm that connected the world long before Zoom or WhatsApp. The company was founded in 2003, and though several countries including Denmark and Sweden claim it as their own due to various links to the company and its founders, Skype was truly the first Estonian unicorn.

To understand how Estonia managed to help create a unicorn tech company only twelve short years after re-independence, we have to start with the story of Jaan Tallinn and Bluemoon Interactive. Jaan was one of the founders of Bluemoon, a gaming company he had created along with several childhood schoolmates.[9] The group behind the company was initially comprised of Jaan, Ahti Heinla, and Priit Kasesalu, all of whom had been involved in programming in their youth. The group eventually expanded to include Toivo Annus, a local project manager who became the glue that held the team together, as well as several other contributors. In 1989 Bluemoon had struck gold, or at least what must have seemed like gold to the young entrepreneurs, with the creation of their first computer game, *Kosmonaut*.

Kosmonaut put the player in the driver's seat of a hovercraft, navigating a range of courses and avoiding obstacles, all while keeping up the oxygen levels in the cabin.[10] It also became the first Estonian video game to be sold abroad. The game raked in the equivalent of $5,000,[11] a king's ransom considering a significant portion of the population got by on less than $10 a day even several years after the collapse of the Soviet Union.[12] The team would jokingly refer to themselves as the "computer game industry of Estonia,"[13] being the main (and possibly only) commercial game developer in the country at the time. Although their first game was successful enough for the group while they were in their youth, by 1999 the industry was becoming more difficult to navigate and Bluemoon was on the cusp of bankruptcy.[14] Jaan and the team would be out of business if they couldn't make some quick cash.

At that critical moment the group came across an ad by Tele2, the first independent phone company in Europe to take on the state-run or formerly state-run telecom monopolies.[15] Tele2 had recently begun operating in Estonia and was looking for programmers to help develop a new site. The EveryDay.com portal was going to be Tele2's answer to AOL, giving customers access to free email and online entertainment as a benefit for subscribing.[16] The product was meant to be a significant differentiator for the company in a newly competitive telecom market and make the company a definitive market leader in Estonia. The Tele2 team leading the development of the portal was composed of Niklas Zennström and Januus Friis. Zennström had joined Tele2 as only its twenty-third employee and Friis, a high-school dropout who had taught himself computer skills while working at the customer help desk

THE BIRTH OF A UNICORN

of an internet service provider,[17] had been brought on to head customer service for Tele2's Danish operations. Both were entrepreneurial to the extreme and passionate about the future of telecommunications and the web, seeing in the nascent digital economy an ability to further disrupt the telecommunications industry. Zennström was especially avid about using new technology to challenge incumbents, saying some years later during a conference on telephony, "Not only is it great fun [to take on monopoly incumbents], it creates huge business opportunities."[18]

For Jaan and the Bluemoon team, surely just as alluring as the potential for disrupting an ancient industry was the fact that the job paid a daily wage that was more than the average Estonian's monthly salary at the time.[19] It proved too good an opportunity not to try for. Even though the project required knowledge of PHP, a programming language that none of the Bluemoon team was particularly familiar with, they managed to successfully complete the test task that was being used to screen applicants,[20] impressing Zennström and Friis. So, the Estonia-based Bluemoon team joined forces with them to develop the portal for Tele2. While the project turned out to be a commercial failure for the company, Zennström and Friis had found the Estonian team to be top-notch engineering talents.

Wanting to fully realize their entrepreneurial inclinations, Zennström and Friis soon left Tele2 after the failure of the EveryDay.com portal. For Zennström, entrepreneurship offered an opportunity to create tremendous wealth, but that wasn't his only motivation. He was also driven to build something new: "Money, for me, was one motivation—but so was the drive to change something, to make something happen. And to prove to the world you can do something real."[21] After considering a range of potential business ideas, the two were inspired by Napster's infamous peer-to-peer (P2P) file-sharing product which allowed a user to easily upload, share, and download files with any other Napster user. This mostly meant that people used the Napster platform to share music online during the days before streaming services like Spotify and Apple Music. However, Zennström and Friis believed that where Napster had failed to work with the entertainment

15

industry they were inadvertently disrupting—leading to the decimation of Napster as a viable business enterprise in the face of a legal and regulatory onslaught—they could build a product that was just as disruptive but that would play nice with legacy media companies.[22] Plus, there was an opportunity to go further than Napster had by making the product easy to use and to enable the sharing of any type of file, whether movies or software programs, tapping into a larger market than Napster's more music-focused product.[23] Thus, the idea for Kazaa was born. And when Zennström and Friis thought about where to recruit the technical talent that they needed to build the product they envisioned, Jaan and the Bluemoon team were the obvious answer.

In 2000, Jaan and the Bluemoon team began the difficult challenge of developing a new technology that would enable fully decentralized P2P file sharing. Unlike Napster, which depended on a centralized system to share the files on the platform, Kazaa would be built in a fully distributed fashion where each user lent their own computer as a node in the network. This decreased costs for the company and made the system significantly more robust.[24] Often working out of his apartment, Jaan and the Bluemoon team developed FastTrack, the protocol which formed the backbone of the unique P2P protocol that powered Kazaa. Meanwhile, Zennström went to Los Angeles to talk to music studios about the potential benefits of the new technology. Although he was armed with a persuasive pitch, proselytizing about the future of the entertainment industry and the need to embrace the future of the web rather than try to fight it, his words fell on deaf ears.[25]

Rather than pivot or create a new product, the team decided to launch Kazaa. They found themselves in court almost immediately. Although a Dutch Supreme Court would eventually rule that Kazaa could not be held liable for files that its users shared over its P2P protocol,[26] the company faced significant legal challenges. In the United States, where the music industry had particularly strong protections, Kazaa faced legal issues which were liable to sink the company and land the founders in significant trouble.[27] While not naïve about the risks posed by launching a Napster-style product,

THE BIRTH OF A UNICORN

the Kazaa team must have been taken aback by the aggressively litigious nature of the industry. In retrospect, however, the numerous lawsuits are unsurprising given that much of the entertainment industry saw the file-sharing technology as an existential threat to their business models rather than the harbinger of an evolution in how consumers interacted with artists and entertainment products.[28] Even with legal challenges and monetization issues, Kazaa quickly became one of the most downloaded products on the web and comprised a significant amount of global internet traffic.[29]

Despite the growth of the business, the lawsuits were a constant irritant. At one point, according to NBC reporting, Zennström totally avoided travel to the United States out of fear of being sued.[30] He was hounded by summons, narrowly avoiding one paired with a bouquet of flowers for his wife and another delivered by motorcycle courier during a visit to London.[31] He avoided air travel, often traveling between Sweden and Estonia on one of the ferries perpetually shuttling between Stockholm, Helsinki, and Tallinn.[32] Zennström and the Kazaa team would eventually decide the stress of avoiding court summons and fearing that every stranger on the corner was a lawyer waiting to slap them with another lawsuit wasn't worth it, especially if it meant the business would never be able to truly thrive.

The company plotted a sale in 2002 that would see Kazaa's brand and platform acquired, but with only a license to the peer-to-peer software that powered it rather than full ownership of the technology.[33] It was an act of tremendous foresight on the part of Zennström and the group, one that would serve them well in the future. Knowing that the file-sharing technology that they had created was incredibly valuable, Zennström, Jaan, and the rest of the team set about thinking of ways that they could leverage it for a new business, hopefully one less legally fraught than the last. According to reporting by Ars Technica, the team played with a range of ideas, including a product that would allow the sharing of home Wi-Fi.[34] But they struck gold when Toivo Annus and Friis pitched an idea that would be a perfect fit for the team's experience in P2P technology and telecommunications. In the original product discussion memo from Toivo to the team, he laid out a grand set of ideas and a pathway to "sweep the carpet from under telcos feet," in ways the big telecommunications companies couldn't even imagine.[35]

What if they could use something like the P2P protocol that they had created for Kazaa which allowed users to share files for free fast and easily around the world, but instead of songs and media, they made expensive long-distance and international calls easy and free? While people under a certain age may consider the idea of paying for long-distance calls foreign, it was a huge revenue driver for major telecommunications companies and in 2001, long-distance calls alone earned American operators billions of dollars in revenue.[36] Telecom companies were also widely considered difficult to deal with for both consumers and other businesses. In tech journalist Kara Swisher's memorable memoir, *Burn Book*, she details a discussion with Steve Jobs who, when asked about the possibility of launching a phone one day, compared telco giants to sphincters given the difficult hoops telecom companies required others to jump through.[37] But with their experience in developing products for a disruptive telecommunications company, building user-friendly P2P infrastructure that had scaled to millions of users worldwide, and working in legal gray zones, the entrepreneurial group was perfectly positioned to develop a new product and disrupt the stodgy incumbent telecommunications industry. The product would be named from of a portmanteau of "sky" and "peer" which was subsequently shortened to "Skyper"—but as the domain name Skyper.com was already in use, it was then shortened further to its final name, Skype.[38]

After a few months of engineering work, the team unveiled Skype to the public in August 2003. The userbase grew incredibly quickly—within two years of its launch there were 50 million global Skype users.[39] This was in part thanks to a unique company culture and a relentless focus on the user experience. According to one employee, "Right from the start we set out to write a program simple enough to be installed and used by a soccer mom with no knowledge of firewalls, IP addresses, or other technological terms."[40] This would be a critical factor in Skype's rise and differentiation from competitors who may have been better resourced but made overly complicated products that placed the onus of figuring out how to work complex software and navigate firewalls on new users. The focus on user experience wasn't the only cultural differentiator for Skype versus other more established players—the early Skype team were a close-knit group who worked hard and partied just as hard. Part of the onboarding ritual for new employees in Tallinn was to go to a bar and take a shot of Millimallikas, a gut-wrenching blend of tequila, sambuca, and Tabasco sauce.[41]

Despite its rapid user growth, the company had a rocky start. This was 2003, only a few years after the massive dot-com crash which had destroyed fortunes (paper or not) and made the venture capital community extremely wary of taking risky bets. Reminiscing nearly a decade later during a lecture at Stanford's Graduate School of Business in the heart of Silicon Valley, Zennström said, "Everyone was still licking their wounds and were very resistant to investing in technology companies."[42] The company was passed over by more than two dozen venture capitalists over the course of a year and a half of attempted fundraising for just 1.5 million euros.[43] Despite its growth, Skype was a risky bet: even though users loved the product, competition was fierce, not just from existing telecommunications companies that were entrenched across the world, but from myriad competitors who were building extremely similar solutions (or at least, who were trying to).[44] Plus, there were still legal issues inherent in a product that enabled low-cost global communications and the innumerable government agencies who wanted to access the contents of the conversations happening on the platform.[45]

Luckily for the Skype team, they found in the Draper family a group of true believers who were happy to invest in a revolutionary idea. The Draper name is well known in the tech and venture capital community. General William Henry Draper, who served the first undersecretary of the army and first U.S. Ambassador to NATO, was the founder of the venture capital firm Draper, Gaither, and Anderson, the first venture capital firm on the West Coast—truly a man of firsts.[46] His son, William (Bill) Henry Draper the Third founded Sutter Hill Ventures before being appointed Chairman of the government run Export-Import Bank, the official export credit agency of the U.S., under President

Reagan. After returning to the private sector in 1994, he founded Draper International, the first U.S.-based venture firm to invest in India.[47] His son, Tim Draper, would become one of the key funders of some of the most ambitious projects of the twenty-first century from Hotmail to Tesla through the venture capital firm DFJ.[48] Even Tim's children are in the family business. Both his son Adam and daughter Jesse are fourth-generation venture capitalists, investing in companies like Coinbase,[49] which operates a platform for buying, selling, and storing crypto assets which went public in 2021 at a market value that briefly exceeded more than $100 billion.[50]

Growing up the son of a venture capital legend, Tim understood both the venture capital industry and that thanks to the internet and increased global connectivity there were opportunities worldwide, even in the former Soviet Union. Tim's father, Bill Draper, had provided $250,000 in seed funding for Skype along with his business partner Howard Hartenbaum,[51] which surely helped Tim and DJF's chances of investing in the company. For Tim and the DFJ team, even though the company wasn't yet profitable, it was clear that Skype was on to something just based on the speed of user adoption. They also knew from the team's work on Kazaa that the technology underpinning the platform would help to quickly differentiate Skype from global competitors, plus the team's focus on user experience meant they had an advantage over companies who hadn't made their products as easy to install and use.

It didn't hurt that DFJ partner Steve Jurvetson, the son of an Estonian refugee,[52] was extremely proud of his heritage and was doubly impressed to see the small Bluemoon team from Estonia building products that even the biggest global companies were having difficulty developing. Jurvetson has developed a reputation for only investing in companies that have a real chance of changing the world, or "ideas that will have entries written about them in history books, if they succeed,"[53] highlighting just how important he believed Skype's technology would be in connecting the world. In an interview with the technology—focused publication Ars Technica years later that dove deep into the founding of Skype, Jurvetson would recall of the Estonian core engineering team, "I remember wondering: how can they be so good?," hypothesizing that such a

THE BIRTH OF A UNICORN

small group could do so much so quickly because of the resource constraints that had been thrust upon them due to the long Soviet occupation which forced the team to program more effectively relative to Western tech companies.[54]

After underwriting the first major funding round for Skype, DFJ ended up with more than 10% ownership of the company according to CNET.[55] For the Skype team, it meant they now had enough money in the bank to scale the company. The floodgates of venture capital opened after DFJ made their initial investment, with investors begging to meet with the company rather than the other way around. With their near-term financial future secure, the Skype team focused on strengthening their product and growing the user base. The company continued to concentrate a significant portion of its software engineering operations in Tallinn. By 2005, the company had grown exponentially and was no longer just fielding inquiries from potential investors, but acquisition offers from a range of suitors. According to Ars Technica,

> In the summer of 2005, Jaan Tallinn was often in London participating in talks with eBay, discussions that were being held at Morgan Stanley investment bank's offices. On one occasion someone jokingly said, "Hey, Jaan, are you gonna sell Skype?" Tallinn replied, "Yes, and I'll be bringing a big suitcase with me to take the money home in."[56]

At the same time, larger and better-resourced competitors had started to fix their flaws and internet giants like Google were launching competing products. In the face of increasing competition, the team decided it was the right time to cash out.[57] In the summer of 2005, they finalized the sale of Skype to another similarly recognizable company of the era, eBay. With an eye-popping price tag of $2.6 billion (a bit more than $4 billion in 2024 dollars adjusted for inflation), Zennström, Friis, and the Bluemoon team found themselves exceedingly wealthy.[58] While the story of Skype after its acquisition by eBay is a long and often contentious one that is shrouded in lawsuits, write-downs, and product integration failures, what was important for the Estonian nation and early employees of the company was its initial meteoric rise and acquisition.

The impact of Skype on the Estonian nation is hard to overstate. It's often credited as the seed that helped to sprout the entire startup ecosystem in the country. According to former early Skype employee Ott Kaukver, "If you look at the startup scene… either you have a Skype investor, you have a Skype founder, or you have people with experience from Skype working [there]."[59] Skype's sale also put cash in the pockets of many of the Estonian founders and early employees and exposed them to the world. The hard data confirms the story—there were about 6 million euros worth of investments in tech companies in Estonia in 2006. Sixteen years later, 1.3 billion euros were invested in the country, a staggering increase.[60]

The Skype "mafia," made up of the Estonian Skype founders and early employees, became prolific in the then nascent local startup scene as both investors and founders. Taavet Hinrikus, Skype's first employee, frequently encountered the arduous task of moving money from euros to pounds and back while working in the London office. From that pain point, TransferWise (now Wise), a leading global cross-border payment transfer company, was born. Like many other companies founded by the Skype "mafia" and Estonian founders, Taavet made sure to leverage Estonian engineering talent, hiring scores of Estonians in key roles, which continued to help build up the local economy and the reputation of Estonia as a hub for top technical talent and entrepreneurship. As a plus, Estonian engineering talent at the time was cheap compared to most global alternatives, especially given how high-quality the talent was thanks to a strong local education system and strong entrepreneurial drive, giving startup founders with local connections a leg up over global competitors in the everlasting hunt for talent.

In addition to being an excellent entrepreneur, Taavet was buoyed by his reputation from Skype, with one venture capitalist who had invested in Skype exclaiming, "I would say Taavet is by far the success story [from Skype]."[61] After raising a whopping $1.7 billion,[62] Wise went public in 2021 (trading as WISE on the GBX stock exchange). As of late 2024, the company is worth about $8.5 billion. Taavet has since become a prolific investor—putting his own money into more than 150 startups across Europe, with around 10% of his personal portfolio allocated to Estonian companies[63]

THE BIRTH OF A UNICORN

while also raising a fund with hundreds of millions of euros in assets under management alongside fellow Estonian entrepreneur and early Skype employee Sten Tamkivi.[64] The impacts of Skype on the Estonian startup ecosystem go well beyond TransferWise. According to Janer Gorohhov, one of the founders of Veriff, an Estonian startup which had raised more than $200 million as of early 2024,

> Skype was the first generation, TransferWise and Pipedrive the second, and now Veriff is part of the third… It's great that we have this network effect that helps the start-up and scale-up scene grow bigger and bigger, and it starts with Skype. That's where so many Estonians learnt how to build startups.[65]

In the early 2000s, running a business was still a foreign concept in much of the former Soviet Union. After decades of occupation with nothing resembling a free market or competitive enterprise, many people simply did not have managerial or entrepreneurial skills and had only heard about the concept of business ownership from older relatives who remembered the pre-occupation days, or from newly accessible foreign media. Skype helped change that for Estonia's tech sector. People like Taavet not only worked at a global company and were exposed to management best practices, but they also helped to build a brand that reached hundreds of millions of people and was instantly recognizable across the world. This experience showed them and the scores of other Estonians who were involved in the company, either before its acquisition by eBay or after its eventual sale to Microsoft, that they could succeed in the tech industry even if few people in Silicon Valley could point Estonia out on a map. According to Taavet, "By creating, having and maintaining Skype in Tallinn, we gained a great insight on how to launch a great global product and it created a feeling that we can create big things in a small place."[66] A kid growing up in Tallinn might physically be very far from the tech ecosystem in Silicon Valley, but they could build a global company from Estonia just the same.

For Jaan Tallinn, as one of the key co-founders and technical architects of the technology underpinning both Skype and Kazaa, the multi-billion-dollar Skype acquisition meant that he had not just time on his hands, but money in his pocket. Rather than buy a yacht or a sports team, he found a much bigger challenge to take on after helping to connect the world. In 2009, he came across the writings of Eliezer Yudkowsky,[67] a prolific writer and philosopher who popularized a new branch of the philosophy of rationalism. Yudkowsky is perhaps best known as a frequent and early voice sounding the alarm about the risks from artificial intelligence (AI) to humanity and as one of the founders and core proponents of the field of AI safety, the study and practice of preventing "runaway" malevolent artificial intelligence from destroying humanity.[68] Think Skynet without Schwarzenegger and cheesy 1980s special effects.

Partially due to Yudkowsky's writings, Jaan became convinced of the importance of AI safety. If he has his way, Jaan's greatest legacy will have nothing to do with Skype but instead it will be for having played a role in saving humanity from extinction. This was a deeply serious turn for a man who had once quipped that he had personally been responsible for saving 1 million human relationships thanks to Skype—but now sometimes added, when speaking about the risk AI poses, that it "doesn't make sense to save human relationships if you don't make sure [people] live longer, and then make sure they don't get destroyed."[69]

Flush with cash from the Skype acquisition, Tallinn began donating to various organizations focused on combating existential risks to humanity[70] from malign artificial intelligence to biological risks like the creation and release of a synthetic virus (a prescient fear given the human and economic toll of COVID-19). However, he became increasingly concerned that humanity was not paying close enough attention to such catastrophic or possibly existential risks and could be sleepwalking (or in the case of some AI startups, consciously running) into a calamity that had the potential to wipe out humanity. He decided to devote much of his life to the field, co-founding a nonprofit research center at the University of Cambridge called the Centre for the Study of Existential Risk in 2012 and the nonprofit Future of Life Institute in 2014, whose mission is to reduce the likelihood of existential risks to humanity. The organization received significant notoriety after releasing a letter in 2023 signed by Elon Musk, Steve Wozniak, and Andrew

THE BIRTH OF A UNICORN

Yang (among others) calling for a six-month pause on the development of cutting-edge AI models.[71]

While the outcome of Jaan's quest to save humanity from ourselves and the technology we might bring into the world is yet to be determined, there is no question that he has left a mark on Estonia. Jaan and the tiny Bluemoon team's peer-to-peer software and superior user experience helped Skype outcompete much larger and better funded companies, putting Estonia on the map as a global leader in tech. Skype's Estonian employees became educated in building global businesses and gained the confidence needed to actually go out and create them, and soon became the next generation of entrepreneurs in the country, helping cement Estonia's place as a global tech leader. And the money that Jaan, Bluemoon, and early Skype employees made would serve as the catalyst for many more entrepreneurs to build their own companies fueled by early-stage investments from the Skype mafia.

In a way, the story of Skype is the story of Estonia. Founded by young, determined, and generally underestimated entrepreneurial-minded people, both Estonia and Skype managed to prevail against all odds. One became a unicorn company, selling for billions of dollars and making its founders exceedingly wealthy while becoming a global brand. The other became a unicorn country, seeding scores of unicorn tech companies from Bolt to Wise, and is globally recognized for its prowess not just in private sector technology but in digitalizing nearly all the country's government operations and both recognizing and leveraging the potential of AI to reimagine government operations well before the advent of ChatGPT. However, to truly understand how Estonia was able to become a global technology powerhouse in a single generation after its independence from the Soviet Union, we have to start with a song.

2

SONIC YOUTH

The haunting song *No Land Is Alone* was created by composer Alo Mattiisen to protest Soviet plans to dramatically expand the mining of phosphorite in Estonia, including in the Virumaa region.[1] Phosphorite mining as practiced by the Soviet Union was deeply destructive, involving strip mining miles of land, decimating the local environment, and the creation of toxic byproducts. The expected waste from new phosphorite mining activities risked the possible pollution of up to 40% of Estonia's water supply.[2] Perhaps more dangerously for the indigenous people of Estonia and their culture, the Soviet plans called for relocating thousands of Russians to work the mines. This was an intentional strategy to bring more people loyal to the Soviet cause to the country and as part of an ongoing policy of Russification to extinguish the culture, language, and unique heritage of the local peoples.[3] For Estonia, which had been suffering under the yoke of Soviet oppression for decades, Moscow's plans to expand phosphorite mining and with it, the continued Russification of Estonia, was the straw that broke the proverbial camel's back.

It was 1987, a time of deep uncertainty in Estonia and the rest of the Soviet Union. Mikhail Gorbachev was part of a new generation of Communist leaders who understood that they had fallen far behind the West in economic development, industrialization, and scientific progress. He knew that if the Soviet system was to have a chance of surviving, fundamental changes were needed. Gorbachev had been a champion of the twin reforms of *glasnost* (openness) and *perestroika* (restructuring) which involved increased freedoms of speech and an embrace of economic reforms since his ascendance to the role of General Secretary in 1985.[4] While Gorbachev's talk of openness and reform were welcomed by many liberal reformers and those in occupied nations, their impact was mixed to start as most were uncertain as to what the reforms meant in practice.

Almost all Estonians had visceral memories of the mass deportations unleashed on their friends, family, and anyone who the Soviets thought might cause trouble or who they felt like deporting to Siberia, and which took place almost as soon as the Soviet Union occupied the country and caused the deaths of thousands. Even former Prime Minister Kaja Kallas[5] remarked in a 2023 *Bloomberg* interview when discussing the scale of deportations and the impact on her family that,

> This is not a unique story in Estonia. I mean, every family has a story like this. My family, my mother was only [a] six-months-old baby when she was deported to Siberia in a cattle wagon with my grandmother and great-grandmother.[6]

Along with (and partially due to) rampant deportations, Estonia and the Baltics had a long history of resistance. "Forest Brothers" melted into the countryside after the renewed Soviet occupation and fought valiantly for decades with the support of their countrymen.[7]

While the Soviet Union was in charge, the KGB could show up at any moment to take anyone they wanted for any reason, or often for no reason at all other than to fulfill a quota and have them sent to Siberia to serve hard labor in virtual exile. In the 1940s alone, tens of thousands of Estonians were deported[8] (a huge number in a country with a population of just over a million people). It was a death sentence for many and a strong deterrent against protest and any activity, however seemingly benign like celebrating Christmas or whistling an Estonian folk tune, that might land someone on a KGB list.

However, the Soviet Union had long been more willing to tolerate discussion on issues that could be considered controversial if they were related to the environment. Marju Lauristin[9] was the daughter of leading Stalinist Soviet-era Estonian politicians. Despite her family background, Lauristin was a champion of Estonian culture and independence, helping to establish the Popular Front in the 1980s as one of the first local political movements allowed during the Soviet Union.[10] With the emerging debate over phosphorite mining in Estonia, Lauristin said that "In the Soviet Union, you could speak about the environment."[11]

When a program detailing the Soviet plans for phosphorite mining aired on *Eesti Televisioon*, the floodgates for popular dissent across the country were opened. Estonians had already been on edge after the 1986 Chernobyl nuclear disaster and were deeply concerned about similar environmental disasters happening in their beloved homeland. Unlike previous disasters in the Soviet Union, Chernobyl had at least been acknowledged by the highest levels of Soviet leadership, with the official newspaper, *Pravda*, reporting on the gross incompetence and cowardice of those involved.[12] Gorbachev later remarked that Chernobyl was one of the events that opened his eyes to the need for major reform in the Soviet Union.[13] Less publicized, but known in Estonia, was the fact that several thousand Estonian men were sent to help clean up the mess in Chernobyl. According to writer and activist Tiit Made[14] in his 2015 book on the history of Estonia's liberation, few of them were given the iodine doses needed to ward off the worst effects of the radiation and most were kept working for nearly double the time they had been told they would be exposed.[15]

Soon, local academics were openly discussing the negative ramifications of the proposed phosphorite mines on the Estonian environment. The consensus, even from institutions that previously had toed the Soviet party line, was that the mines could very well end up an environmental disaster. And when students in Tartu, Estonia's intellectual capital, and Tallinn, Estonia's actual capital, began debating and protesting the plans, they were allowed to do so with minimal consequences.[16] The lack of Soviet action on the burgeoning movement helped foster further debate and discussion.

29

Seeing the unexpectedly forceful protest from broad swaths of Estonian society, local Soviet apparatchiks quietly shelved the plans for the mining expansion in the hope of placating the populace.[17]

Instead, the people were emboldened and the conversation about mining and the environmental degradation of Estonian land under Soviet occupation began to transform into a larger discussion about the past and future of the Estonian people. Soon, things moved even faster in the burgeoning Estonian independence movement. Mart Laar,[18] the political activist, historian, and future Prime Minister, would later say, "Between August 1987 and February 1988, the atmosphere in Estonia changed more than it had during all of the preceding 40 years. People awoke to the possibilities of public protests."[19] More protests were organized and activist leaders like Trivimi Velliste and Heinz Valk began discussing the potential of Estonian independence. Some statements like the famous September 1987 article "A Proposal for Full Economic Self-Administration of the Estonian SSSR" by political leaders Siim Kallas,[20] Tiit Made, Edgar Savisaar,[21] and Miik Titma[22] were more veiled than others in their attempts to move Estonia towards sovereign nation status. The strategies between political activists differed, but all played a role in the independence effort.[23]

Public protests, various op-eds, and debates over economic and cultural sovereignty—as well as the anti-phosphorite campaign—were critical in Estonia's path towards independence. But one uniquely Estonian cultural phenomenon had a great deal of impact. Estonia (or the land that is now the nation of Estonia) has been inhabited for many centuries. During that time, despite repeated invasions and occupations, the country has developed a unique local culture—one built in no small part around song. Over those years of history, Estonia has amassed one of the largest collections of folk songs in the world[24] and has a deep tradition of using music to gather the Estonian people together and bind them in a shared experience. The use of song even extends to a form of battle, with author Anatol Lieven remarking that, "both Estonian and Finnish folklore record traditions of 'singing matches', a kind of peaceful single combat between rivals."[25] Gathering Estonians and creating shared cultural experiences often happens through national song festivals. These festivals bring together hundreds of thousands of Estonians from across the country (and in post-Soviet years, people from across the world), an impressive feat.

However, the song festivals, especially during the Soviet occupation, were not always joyous affairs, reflecting the darkness of Estonia's recent history. The early festivals after occupation were especially emblematic of the cultural, political, and social destruction wrought by the Soviets, when average citizens were arrested on the festival grounds and as propagandists sought to co-opt the usually nationalist event to advance the Soviet worldview with pro-Communist songs. Despite repeated attempts by the Soviets to turn what had started as a pathway to developing an Estonian national identity and twist it into a Soviet propaganda festival, the event remained important for the Estonian people. And though the Soviets went so far as to imprison attendees and composers who failed to toe the party line, it managed to retain a local cultural element.[26]

Perhaps the boldest of such Soviet attempts to usurp the identity and purpose of the festival was during the widely boycotted 1980 Moscow Olympic Games when Tallinn hosted the sailing competitions. The song festival that year was made to be integrated into the cultural program for the games and, likely out of fear of being embarrassed on the world stage, authorities increased pressure on potential dissidents. According to journalist Maris Hellrand, this led, at least in part, to the emigration of famed Estonian composer Arvo Pärt[27] who was already under fire from the authorities for the religious aspects of some of his compositions.[28]

The heavy-handed Soviet coercion tactics could not stop the fierce patriotic undertones of the song festivals. Whether it was the wearing of traditional clothing, composers slipping pro-nationalist lyrics past censors, or the spontaneous singing by the audience of folk music frowned upon by the Soviets, Estonians generally maintained the festivals as their own and as a place of national identity. According to Marge Allandi, who studied the history of the festivals, they create a unique atmosphere of unity and belonging.[29] And as they did not lose their local cultural relevance despite the best efforts of Soviet propagandists to usurp them, the festivals and tradition of song became critical in the nascent independence movement ignited by the earlier campaign against the phosphorite mine expansion. This sense of unity would become even more important after the country regained independence as political leaders embarked on a series of ambitious reforms to pull the country out of the economic doldrums created by Communist rule.

The importance of song to the independence movement was most obviously manifested in the summer of 1988, shortly after a protest in Tallinn's Hirve Park marked the first time an open and free political demonstration discussed the infamous Molotov-Ribbentrop Pact during the Soviet occupation.[30] The local tourism industry, even during the Soviet era, has long taken advantage of the pristine city center of the capital. Beginning in the early 1980s the city started what is now an annual festival called Tallinn Old Town Days. During the 1988 festival, a range of musical acts came to perform including from neighboring Finland. In a burst of spontaneity to please the crowds full of patriotic fervor who were eager for nationalistic music at the end of the planned events on the last day, the bands and their audiences marched from the square in the old town to the Song Festival Grounds and continued to sing.[31]

From there, the event turned into something much more than a concert: it became a sonic manifestation of a people's desire for freedom. According to some estimates, the festival hosted tens of thousands of attendees and even the Soviet authorities present dared not intervene and repress the demonstrations.[32] That demonstration of national unity, along with the broader independence movement, became known as The Singing Revolution. Artist and activist Heinz Valk coined the name in an op-ed soon after the festival to capture the unique nature of Estonia's push for freedom.[33] He would later say, "Until now, revolutions have been filled with destruction, burning, killing and hate, but we started our revolution with a smile and a song."[34]

There was soon a new opportunity for the Estonian independence movement to act on a grand scale: the August 1988 national Song Festival. The event was slated to be one of the first Song Festivals that would be held under Soviet occupation but without a significant element of propaganda (thanks to earlier activism and the loosening grip of the Communist Party). A range of speakers were planned, almost all Estonian nationalists of varying degrees who wanted, at a minimum, more autonomy for their country. The songs were made up of traditional and patriotic folk classics while the Estonian attendees waved their country's previously banned flag freely (many of the flags had been hidden away deep in cellars, attics, or buried in gardens during the Soviet occupation in the hope that one day they could be flown again).[35]

Although the Popular Front organizers had planned to take a more conciliatory tone and try to avoid the ire of the KGB officers in attendance and the politicians in Moscow, it was clear that attendees were not going to be easily controlled. That included their own planned speakers. When Trivimi Velliste, a well-known Estonian political activist, took the stage, he had some prepared remarks but mostly spoke from the heart. While the Popular Front organizers had expected him to take a more nuanced tone, his speech was anything but: he demanded in front of the audience of hundreds of thousands that Estonia break away from the Soviet Union and restore its independence. His remarks were received with elation by the attendees.[36]

There were more than six hours of speeches throughout the day of the festival with numerous firsts and notable moments. One of the most well remembered was a statement by Heinz Valk, who had coined the phrase "The Singing Revolution" a short time earlier. At the opening of the event to a crowd representing somewhere between 10% and 20% of the population of the entire country, Valk said the immortal words, "One day, no matter what, we will win!"[37] Little did Valk or any of the other attendees know just how soon those words would come true.[38] The rousing speech was emblematic of the Estonian mentality—that no matter the cost, no matter the hardship, they would do what was necessary to throw off the shackles of Soviet oppression and create a successful country that would never be subjugated again.

Even with the tremendous momentum behind the popular protests and independence movement, the path towards independence was not clear and another bold statement was needed to help usher independence and break the Soviet stranglehold on the country.

The fiftieth anniversary of the secret Molotov-Ribbentrop Pact, which had led to the annexation of much of Eastern Europe, including the Baltic nations of Estonia, Latvia, and Lithuania, and served as a potent symbol across the world of the evils of totalitarianism, would be on 23 August 1989. Hoping to bring renewed international attention to the Estonian independence movement, it has been reported that Estonian politician Edgar Savisaar proposed a unique concept for a nonviolent protest event at a joint meeting between democratic reform parties from Estonia, Latvia, and Lithuania (although the claim as to who originated the concept is somewhat disputed).[39] The groups needed to demonstrate Baltic unity and their desire for freedom and self-rule to the world. Especially important was the need to demonstrate their resolve to key Western leaders, many of whom were more interested in propping up Gorbachev and his relatively moderate administration than pushing for the independence of the tiny Baltic countries.[40]

The idea would become the Baltic Way (sometimes called the Human Chain), a single large-scale nonviolent protest which would see millions of people link hands across a span of hundreds of miles to create a tangible sign of their unity and desire for freedom. Most would have shrugged off the protest as overly ambitious even if they had years to plan for it and a tremendous budget, but the independence advocates were undeterred. Prime Minister Laar, speaking later about the many incredible feats accomplished by early political leaders in Estonia, explained that "we did not know what was possible and what was not—so we did impossible things."[41]

The scale of the event must have astounded and frightened the Soviet authorities. They reacted to the news of the forthcoming protest by trying to scare away potential attendees while casting aspersions on the organizers and the basis of the event. Eventually, desperate to stop the protest and the international incident it was likely to catalyze, the Communist Party tried a last-ditch attempt to appease the Baltic people. According to *Washington Post* reporting at the time, the chair of a commission set up by the highest state authority in the Soviet Union to "investigate" the Molotov-Ribbentrop Pact admitted that the previously secret agreement was real, but in the same breath the commission claimed nothing had changed regarding Eastern European states' incorporation into the Soviet Union from a legal perspective.[42]

The Soviet ploy backfired spectacularly, leading to international coverage of both the announcement as to the existence of the pact and of the coming protest. Still, no one was sure whether the organizers' grand vision of a human chain across hundreds of kilometers and multiple nations could be achieved. But on 23 August 1989, what many believed to be impossible was manifested for the world to see. Stretching from Tallinn to Vilnius, almost 2 million people joined hands,[43] linking one Baltic capital to the next for over 600 kilometers (approximately 375 miles or roughly the distance from Boston to Baltimore or Amsterdam to Berlin).

The effect that the event had on the Estonian people and the world was tremendous. To this day, many people who joined hands consider it one of their most important memories. In an interview with NPR, Inita Dzene, who attended the event as a teenager, said through a translator:

> Everyone joined hands, she says. She felt the energy of 2 million people flow through her body. That epic demonstration was followed by mass protests in East Germany and Czechoslovakia, and then the fall of the Berlin Wall and independence for the Baltic nations. The infamous Hitler-Stalin pact seemed like history.[44]

Even those who weren't there physically were moved. Kaarel Piirimäe, a historian at the University of Tartu, made an unfortunate but telling comparison in an interview thirty years later that: "While he didn't take part in the human chain himself, he said he can remember the day. 'It's like 9/11,' he said. 'Everyone knows where they were when it took place.'"[45]

The Human Chain sent a powerful message to Moscow and to the world regarding Baltic unity and a commitment to achieving independence through nonviolent means. In 1989, only two years after the fateful phosphorite protests, it was still unclear whether independence was achievable, but what was clear was that the people under Soviet rule yearned for autonomy and a return to independence. For the people of Estonia, the events helped to re-solidify a sense of national identity and common cause, revitalizing the shared culture that the Soviet occupiers had tried to snuff out. This would soon enable the nation to come together and work towards ambitious goals that would have sown irreparable divisions in a less cohesive society. The Human Chain showed the world that Estonia, along with its Baltic neighbors, would take a path characterized by nonviolence and national solidarity towards freedom and democratic governance. As Heinz Valk had christened it, theirs would be a Singing Revolution.

3

INDEPENDENCE DAY

Coups rarely turn out well, but the failed 1991 coup in the Soviet Union was an exception for then occupied Estonia. In August 1991, three years after the events of the Baltic Chain and despite the constant agitation of the Estonian people for independence, negotiations with the Soviet Union for their freedom were moving slowly. The Estonians had little leverage with thousands of Soviet troops stationed throughout the country. With them, the threat of force which could be used to violently suppress any protest movement or extradite any politician, student, or even child to Siberia remained. Despite the dangers, independence movement leaders were relentless in trying to negotiate a peaceful transfer of power for a new and independent Estonia.

Still, with those efforts leading to little practical improvement for the average person and meaningful change, there was concern that it would all be for nothing. In his book *Estonians' Liberation Way*, Tiit Made recounted how just a day before the coup attempt there was no clear warning that the Soviet system was about to collapse.[1] After the Singing Revolution and numerous efforts to establish full independence, Baltic leaders were still negotiating with Gorbachev in an attempt to stave off bloodshed and find a peaceful solution. Estonia had been lucky so far; despite years of popular protests and increasingly bold attempts at domestic reform without the explicit agreement of USSR authorities, there had been no new mass deportations or significant violent conflict between protestors and police forces.

Nearby Lithuania had not been so fortunate. The Baltic country was the first republic to formally attempt to fully break away from the Soviet Union on 11 March 1990.[2] The country decreed the restoration of the pre-World War II government, invoking international law which had long recognized the Soviet occupation of the Baltics in 1940 as unlawful.[3] The move triggered an opening salvo from Moscow in the form of economic warfare in an effort to coerce the Lithuanian people by starving them of food and fuel.[4]

With Lithuanian popular support continuing to coalesce in favor of independence despite (or possibly due to) the Soviet use of non-kinetic measures to influence ongoing negotiations, and with significant media coverage of the event in major Western papers like *The New York Times* and *The Washington Post* which mentioned the issue in 623 and 358 articles respectively,[5] Moscow decided to take drastic steps to counter the independence movement. In January 1991, Soviet troops in Lithuania's capital, Vilnius, undertook a mission to capture key pieces of infrastructure. This included the all-important TV tower which hosted critical communications facilities.

Rushing to ward off the invaders, unarmed Lithuanians attempted to block Soviet troops from entering the facilities. According to on the ground reporting by *The Guardian* in Vilnius, a local police officer with only a nightstick tried to hold off the soldiers by appealing to their humanity, saying, "You're human beings, don't shoot," before being silenced by a soldier who struck the officer in the face with the butt of his weapon.[6] According to eyewitness accounts, at least one person was crushed by a tank before the crowd dispersed.[7] By the end of the day, fourteen had been killed and even Gorbachev was reportedly alarmed at the violence that had taken place.[8] Although there is some dispute as to whether Gorbachev had authorized the use of force and was aware of what was likely to happen, the violence clearly ran counter to the principles of *glasnost* and *perestroika*.[9]

The events in Lithuania led to increasing divisions between Soviet and Russian political actors, namely between General

INDEPENDENCE DAY

Secretary of the Communist Party of the Soviet Union Mikhail Gorbachev, Boris Yeltsin, who effectively served as President for the not yet formally independent nation of Russia,[10] and a number of USSR hardliners from various political factions and security apparatuses, who were committed to maintaining the status quo and keeping the Soviet Union together, by force if necessary. According to Paul Goble, an analyst and special advisor to U.S. Secretary of State James Baker who later assessed the importance of the events for the Jamestown Foundation,

> Boris Yeltsin immediately flew to Tallinn, met with Estonian and Latvian leaders (the Lithuanians could not make it there), and then issued a joint statement recognizing the sovereignty of all three Baltic countries. On his own, Yeltsin called on Russians serving in the Soviet forces not to obey illegal orders to fire on citizens exercising their rights to demonstrate and seek independence.[11]

This unprecedented step by Yeltsin provided a major show of support for the Baltic independence movements from one of the most powerful political actors in Russia, who also had significant influence over decision-making in the USSR.

Where the breakup of the Soviet Union had previously been unthinkable, the threads holding the USSR together were quickly unraveling. Despite the increasing willingness to circumvent the Soviet system by Estonian, Russian, and other political leaders, at the time Gorbachev was still nominally in charge. The fact that the Soviet Union controlled thousands of Russian troops in Estonian territory and was Estonia's main trading partner made his support for independence a necessary precondition for any enduring change (at least while the USSR still existed).

Seeing the massive popular support for independence in the Baltics and an increasing unwillingness by military and political leaders to use force to stymie popular protest along with myriad domestic political and economic factors at play, Gorbachev began working with the Soviet republics on a New Union Treaty. The treaty would remake the Soviet Union into a "Union of Sovereign States,"[12] which was intended to be signed and brought into force by August 1991. The treaty would have seen a dramatic shift in

Soviet policies. According to reporting at the time by Pulitzer Prize-winning journalist and author David Remnick,

> Under the new treaty, the central government in Moscow would continue to control the state's military and security organs, formulate foreign policy, organize the financial and credit systems and control gold, energy reserves and other resources it deems necessary.[13]

Gorbachev asserted that it would be the key in stabilizing the discord across the Soviet republics and was his attempt to prevent the Soviet Union from falling apart, which he claimed would end in "bloodshed."[14] The deal would allow each republic broad latitude in managing property rights and how they chose to run their economy, an unprecedented compromise that was clearly an acknowledgement of the significant failures of the Soviet economic system.

The New Union Treaty was rejected by the Baltic states for not going far enough towards the independence which their people demanded. On the other hand, Soviet hardliners saw it as a step too far, heralding the end of the USSR and emblematic of the perceived failures of *glasnost* and *perestroika*. On 18 August, just two days before the planned signing of the New Union Treaty, the unthinkable happened. Gorbachev, who was at his dacha in the Crimean Black Sea, was accosted by several members of his own senior staff and urged to impose a nationwide state of emergency. He refused, leaving the coup plotters to remit him to house arrest and cut off his access to the outside world. At the same time, troops nominally controlled by hardliners in the USSR poured into Moscow and attempted to seize key government buildings and infrastructure. The state television operator, rather than reporting on the coup, aired Tchaikovsky's ballet *Swan Lake*. This was certainly interpreted by most Soviet citizens as portending bad news as the same ballet had previously aired after the death of several Soviet leaders as an ineffectual distraction while senior political leadership decided what to tell the public.[15]

According to Reuters, the coup leaders, including the Soviet Vice President, KGB Chairman, Defense Minister, and Interior

INDEPENDENCE DAY

Minister, soon issued a statement which claimed that "they were saving the Soviet Union from a 'national catastrophe' and Gorbachev was now 'resting' to get his health back."[16] While Gorbachev was still trapped in Crimea, thousands of Russians who had embraced increased openness and economic reform took to the streets in protest. Leading the charge was Boris Yeltsin, the same Russian politician who had previously signed the treaty of mutual recognition with the Estonian government.

At great personal risk, Yeltsin went to the government office for the Russian Soviet Federative Socialist Republic, which he led at the time, and confronted the soldiers now stationed outside by the coup plotters. Looking to address the crowd of protestors and head off any violence between them and the soldiers, he climbed atop one of the tanks positioned outside the building. Speaking to the crowd, which now numbered in the thousands and included soldiers who had joined the popular protest, Yeltsin proclaimed that all actions by the putschists were unlawful and demanded Gorbachev's freedom and an end to the coup.[17]

According to Yuri Ivanilov, a Yeltsin supporter who was present at the time, "There were soldiers who would have carried out an order to shoot, but nobody was willing to take the terrible responsibility of issuing such an order... You can't organize a coup by committee."[18] After Yeltsin's rousing speech and with no direction from the group of putschists, the coup was effectively at an end. It took another two days for it to completely fizzle out and see the plotters arrested with Gorbachev only returning to Moscow on 22 August. While his return was certainly a relief both for himself and the reformers across the Soviet Union, Gorbachev's standing as a political figure was hugely diminished relative to the now ascendent Yeltsin who had essentially saved Gorbachev's life. The coup meant that the Iron Curtain had been pulled open for all to see the dysfunction of the Soviet system.[19]

For the Estonian people, the chaos of the attempted coup by Soviet hardliners was the opportunity that the ringleaders of the independence movement needed to finally take back their country. Heinz Valk, whose legendary turn of phrase had become the calling card of the singing revolution, said that "the coup in Moscow [gave] us a chance comparable to that in 1918."²⁰ On 20 August, the first day of the attempted coup, the putschists in Moscow had ordered troops to gain control over infrastructure like the Tallinn TV Tower in much the same way they had violently attempted to take over media and communications facilities in Vilnius.²¹ However, this time, they were doing it under the scrutiny of foreign journalists who had been sent to the Baltic capitals after the events in Vilnius became global news.²² Understanding the importance of the moment and the potential implications for the country, Estonian politicians gathered in an emergency session. Quickly, a majority of the members of the Supreme Council of the Republic of Estonia voted to approve a declaration of Estonian national independence, declaring Estonia to be a free and sovereign nation.²³

The Tallinn TV Tower, which at a staggering 313 meters is visible from miles away on a clear day, served as a nerve center of communications infrastructure, much like the Vilnius TV Tower. With its usefulness for communications came military and political importance as the Soviets understood how critical being able to control key communications channels to the populace was in the fight for independence. The Estonian people also certainly understood the value of being able to speak directly to the average citizen—not only because they had seen how hard the Soviet Union worked to block access to foreign media broadcasts but because of the cultural impact that television signals from nearby Helsinki had on the country. For many, Finnish broadcasts gave a glimpse of life outside the Iron Curtain and fostered a pro-Western bent in those who were able to tune in to Finnish TV and reruns of American TV shows like *Dallas* and *Dynasty*.²⁴ Ordered to secure the tower, Soviet soldiers attempted to breach the facility. The outgunned Estonian defenders of the tower had little chance of stopping the Soviets, and soon the soldiers penetrated the TV tower up to the twenty-second floor, which held the most critical communications facilities.

Four men, Peeter Milli, Jaanus Kokk, Uno Kasevāli, and Jüri Joost, barred the entrance to the twenty-second floor in the hopes of buying time. They surely knew that in Lithuania, many unarmed civilians had been killed just a few months prior when attempting to defend their TV tower, but still they decided that the fight for freedom was worth the risk. Recalling the events of that day, Joost said, "I never believed that we would come down from the TV tower on our own feet."[25] While they remained barricaded, holding out against Soviet forces, they were able to keep the broadcasts playing and alert the nation to everything happening in both Tallinn and Moscow.[26]

Throughout 21 August, as events were still unfolding in Moscow and the coup was quickly falling apart, Soviet forces attempted to negotiate with the Tallinn TV Tower defenders. It was increasingly clear that in Tallinn and across the quickly crumbling USSR, soldiers were unwilling to commit acts of violence, especially against unarmed peaceful protestors. Whether this was out of fear of later prosecution and arrest in these countries, out of moral righteousness, due to loyalty to political actors like Yeltsin who had previously urged Russian soldiers stationed in the Baltics not to harm civilians,[27] or simply because in a hierarchical system there had been a breakdown in leadership at the top due to the coup is unclear.

The standoff continued throughout the day and into the next. As the hours ticked by, it became increasingly obvious that the coup was failing, and that Yeltsin was in control in Moscow along with other democratic reformers. Even in far-off Washington it was becoming clear that the coup was unlikely to succeed, as documented in Una Bergmane's *Politics of Uncertainty: The United States, the Baltic Question, and the Collapse of the Soviet Union*. "According to historian Serhii Plokhy, a document that arrived at the White House during the day [19 August] and a CIA report that was presented at a White House meeting convinced the administration that the coup was likely to fail."[28] By early evening on 22 August when news of the arrests of the putschists was heard in Tallinn, the threats from the soldiers ceased as it became clear their previous orders were now to be considered unlawful.

As the siege abated the defenders exited the tower. The Soviet troops were still below, but rather than facing animosity, the Estonians were greeted warmly. Peeter Milli, one of the Estonian tower defenders, later said "The Russian soldiers were all right in the sense that after all was over they shook our hands... We all had orders from our superiors."²⁹ It is impossible to know what would have happened in Estonia if the tower had been captured and the country had been cut off from vital information about the coup attempt in Moscow, but clearly those who defended the tower with their lives engaged in heroic action for their country.

And indeed, it was now their country. While Estonia had declared independence two days earlier on 20 August, it wasn't until the 22nd that the first foreign nation acknowledged that independence. The usually unsentimental Estonian people later named the Tallinn street where the Ministry of Affairs is based *Islandi väljak* [Iceland Square] in gratitude to the Icelandic people who took the first step in acknowledging Estonia's newfound independence.[30] Immediately following Iceland's recognition, many of the Nordic countries, with which Estonia had long enjoyed close relations, either formally recognized the newly and once again independent nation of Estonia or, in a nod to the long-fought battle over the legitimacy of Soviet rule, simply re-established their diplomatic relations with the country after the multi-decade pause that took place during the illegal Soviet occupation.[31]

While crumbling, the Soviet Union was still powerful and dangerous for small Estonia. The country shared a large land border with Russia and was only a few hours away by car or ship from the imperial hub of St. Petersburg. With thousands of Soviet troops who technically still took their orders from Gorbachev stationed across the country, getting official recognition from the USSR was an important milestone for Rüütel, Meri, and other political leaders of the time. Thankfully, on 6 September 1991, the USSR State Council recognized Estonia's independence. Although it would only be a few months before the Soviet Union ceased to exist, achieving independence from the entity which had subjugated it for decades meant that the Estonian people were truly free. They had their own nation once again. But now they had to undo the years of damage from Soviet rule and chart their own path into the future. The crucible of occupation had forged a fierce people who would not only put their lives on the line for their freedom and values, but who now had a reinvigorated culture and were ready

INDEPENDENCE DAY

to embark on a mission to create a new state fit for the modern era. The road to building a new nation would not be easy, but the Estonian people were unified in the need to rebuild in a way that would help them leapfrog both friends and enemies so that their nation would never be occupied again.

4

A NATION REFORMED

The situation for the fledgling Estonian state post-re-independence was desperate. Setting up the infrastructure necessary to deliver services for a new country is hard. Doing so with gangland-style violence in the capital, an economy in chaos, an occupying force numbering in the thousands still stationed across the country, and the remnants of a corrupt and sclerotic Soviet government bureaucracy is something else entirely. For those who know the Tallinn of recent years, it is difficult to imagine the placid capital as a haven for illicit activity, but in the early 1990s it was one of the most dangerous cities in the world.

According to a 2012 study, the homicide rate in 1994 was 29.1/100,000 inhabitants[1]—significantly higher than western European rates at the time[2] and around that of Mexico's 2021 homicide rate.[3] Brutal attacks in Tallinn were common as gangsters tried to gain control of territory and lucrative newly privatized enterprises. The number of reported crimes spiked more than 250% between 1987 and 1997, with most cases being related to theft.[4] The capital city with its port, accessible throughout the winter, had always been a coveted possession. Tallinn has served as a regional trade hub since at least the days of the Hanseatic League, a merchant trading organization operating throughout Northern Europe beginning in the 1300s.[5] With the Soviet Union gone and regional rule of law increasingly fraught, "entrepreneurial" individuals began to see Estonia, and in particular Tallinn, as a newfound opportunity to build wealth, either by capturing actively privatizing local industries or by using the port to export goods from other parts of the former Soviet Union.

The proverbial gold rush, perhaps more accurately described as a regional looting spree, helped earn Tallinn the moniker "the Wild East."[6] It was also nicknamed "Metallinn" as Estonia briefly became one of the largest exporters of non-ferrous metals like copper, brass, and tin in the world.[7] Achieving such a goal would be quite a feat for any small nation, but the fact that Estonia had basically no such domestic metal deposits made the achievement all the more impressive.

In those heady days, the violence between various criminal enterprises and against ordinary people caught in the crossfire became so savage that many saw the gangs as more than petty criminals and a security threat to the nation.[8] One group of ordinary Estonians which formed to fight against the scourge of organized crime managed to get hold of a small former Russian warship as well as a variety of Soviet-era weaponry but answered only to themselves, becoming "vigilantes like in an American movie."[9] Lagle Parek,[10] a dissident during the Soviet occupation who was charged with maintaining domestic security as Minister of the Interior from 1992 to 1993, later recalled,

> Things were really bad. A crime wave from Russia rolled over Estonia and on to the West. Stealing metal, robberies, a lot of homicides. The prisons were so full. Sleeping nets hung under the ceiling and the floors were full. When I was a prisoner, they were never that full.[11]

The decline of rule of law and the desperate economic situation for many ordinary people in the region meant that thefts of anything of value, especially easily tradable commodities, became increasingly common. Estonia was not unique in this: the country's Baltic neighbors struggled with the same issues. Thefts ranged from the sacrilegious in the form of pilfered graveyard crosses[12] to the tragicomic with the theft of a bronze plaque from the door of the parliament building in newly independent Riga.[13] The illicit metals trade was so brazen and widespread that it affected electrical equipment and railway lines as people pried up cables to harvest the copper wire and stole parts of train tracks, further degrading the already poor infrastructure.[14]

Estonia's new leaders were grappling with many competing crises, but none so critical as the dire economic situation. When people are desperate enough to sell graveyard crosses, it serves as a call for unprecedented reform. While Estonia has undergone many ambitious policy initiatives post-independence, one of the most impactful was the introduction of the Kroon. While occupied, Estonia had been incorporated into the Soviet Union's monetary system and forced to use the Ruble, but after achieving their hard-won freedom the Estonian people were eager to assert their independence. One of the most tangible ways to do so, and one which Estonian economists and politicians like Mart Laar believed would help further decouple Estonia from the unstable Russian economy, would be the creation of a local currency.

According to Laar, "Estonia was absolutely dependent on Russia, which accounted for 92 percent of Estonian international trade."[15] That level of dependence would be dangerous for any country but was significantly more so for a small nation which had only just regained independence from the revanchist power which still had troops stationed in the country. And while the Ruble had served as something of a stable currency during the heyday of the Soviet Union, as the former Soviet Republics opened to global markets the Ruble quickly underwent a period of runaway inflation. Despite the clear need for an alternative to the Ruble, Western advisors and global financial institutions like the IMF and World Bank were deeply opposed to the Baltic countries creating their own local currencies, believing that the local central banks lacked the capacity and experience to fully manage their own monetary policy. A 1992 IMF report focused on Latvia stated that, "The introduction of the new currency, the Lat, should be delayed. Little progress has been made in developing the central bank's ability to control monetary policy."[16] For these global financial institutions, the situation in Estonia was not much different, but local political leaders like former Prime

Minister Tiit Vahi,[17] under whose administration the Kroon was launched, and the then leader of the central bank, Siim Kallas, doggedly worked towards the development of the currency.

By the summer of 1992, only a year after independence, the Estonian Kroon was launched to significant domestic fanfare, apprehension among global financial institutions, and regional jealousy. According to author Anatol Lieven,

> When the Estonian Kroon (Crown), the first new currency in the former Soviet Union, was finally introduced in June 1992, it was an object of deep envy for the other two Baltic States. Its re-emergence was a source of great pride to the Estonians and considerable prestige internationally.[18]

The Kroon also helped to solve one of Estonia's most pressing national security threats: the thousands of Russian soldiers stationed in their country. Due to the decline of the Ruble, Russian troops faced increased costs for even basic supplies, with one officer telling Lieven that "after the introduction of the Estonian currency the army could simply no longer afford to feed itself," which surely contributed to Russia's willingness, albeit begrudgingly, to remove troops from the country (in addition to significant Western diplomatic pressure and economic coercion).[19]

Despite the initial fears of international advisors, the Kroon performed well. Soon the IMF was praising the work of the Estonian state bank which had tirelessly pushed for the launch of the currency against foreign advice.[20] Rather than depending on the playbooks of foreign donors and international institutions, which often recommended strategies that made little sense based on the local context, Estonian leaders like Mart Laar, Siim Kallas, and scores of others were willing to forge their own path when it mattered most, spurning foreign experts whose advice was sometimes derided behind closed doors. There was even a joke that made the rounds that Western experts "have never met a pig, let alone a post-Soviet pig" as many of these advisors were happy to provide recommendations to the important agricultural sector based on experiences in other countries with vastly different cultures and problems, rather than working to understand the challenges of the local environment.[21]

A NATION REFORMED

The launch of the Kroon has been hailed as a central tenet of Estonia's successful post-Soviet economic reform. Not only did it show that the country had the state capacity to manage the introduction of a new currency shortly after independence in extremely difficult regional economic conditions, but it also allowed Estonia a measure of true independence. While significant amounts of Estonian trade, from energy to manufactured goods, may have still been with Russia, by decoupling from the Ruble, the Estonian government could now develop a truly independent monetary policy. This newfound freedom would allow them to use novel financial tools to deal with the inefficiencies from the poorly functioning Soviet manufacturing sector and an agricultural industry still reeling from forced collectivization policies.

The Kroon was a significant part of Estonia's successful reform and transition to a leading market-based economy, but it is only part of the story. Less well known is the role of Mart Laar, the then thirty-two-year-old Prime Minister who was elected in September of 1992 with a mandate to truly break away from the Soviet past and lead Estonia into the future. Laar comes across as a gregarious and plainspoken figure often pictured sporting a mischievous grin. Trained as a historian, he is a prolific writer on Estonian history. *War in the Woods: Estonia's Struggle for Survival* is a deeply researched monograph about Estonian resistance during Soviet occupation which Laar published in 1992, risking his life by researching such a contentious topic prior to independence.[22] But while he was a history expert, Laar was significantly less well versed in economics. Before beginning his term in office, he had proudly only read one book on the subject, the free-market screed *Free to Choose* by Milton and Rose Friedman, not including Marxist books on economics which he claimed, "don't really count because they're all wrong."[23]

The Friedmans' words could not have found a more receptive audience. More than a decade later when accepting the Milton Friedman Prize for Advancing Liberty, Laar recalled that:

> Estonia got a lot of advice from other nations about how to work toward freedom. A lot of western countries, including the United States, gave us advice supporting a big state, big government, big expenditures, high taxes, and progressive taxation. And in this

context, I must say it was very useful, again, to remember the Soviet time. Because the first time I heard the name Milton Friedman was in the deep Soviet time, when I read in the newspapers or in some propaganda newsletters of a very bad, dangerous western economist called Milton Friedman. At the time, I didn't know anything about Friedman's ideas, but I was quite sure if they were so dangerous to communists, he must be a good man.[24]

The free-market principles espoused by Friedman found a willing acolyte in the newly anointed prime minister. Knowing the depth of the challenge that he faced in reforming an economy in crisis, Laar turned to those whose ideals mirrored or were highly aligned with those of Friedman and his own. Their task would be to develop a strategy to help the Estonian economy not just to escape the Soviet system that had dragged it down, but one that would help supercharge productivity in an attempt to catch up to the GDP per capita and quality of life found in neighboring Nordic nations which were often held up as a benchmark by Estonian society. Several of the main contributors to Laar's ambitious plan included the Heritage Foundation and the Adam Smith Institute, two free-market-focused American think tanks, along with the International Republican Institute, a nonpartisan NGO with a mission of promoting democracy globally, and Sweden's free-market foundation, Timbro.[25] According to a paper co-authored by Rainer Kattel, Professor of Innovation and Public Governance at University College London's Institute for Innovation and Public Purpose and one of the foremost experts on modern Estonia's development, "the country was now a disciple of Adam Smith's famed 'invisible hand.'"[26] Given both Laar's and his inner circle's proclivity towards free market ideals and the allergic reaction of Estonian officials to anything resembling communist or socialist economic policies, it is no surprise what happened next.

The government soon introduced a series of reforms focused on making Estonia a free-market paradise and eliminating potential sources of corruption. Export restrictions were abolished. Subsidies were removed and price controls on goods were eliminated. A law was passed so that only balanced budgets could be presented to the parliament, codifying fiscal responsibility into statute. The importance of codifying critical policies so that successor politicians in future generations respect the rules created and are prevented from succumbing to corruption has been highlighted as being of critical importance in creating a welcoming regulatory environment for startups and innovation by investor Steve Jurvetson.[27] Creating such a stable legal environment not prone to political disruption has long been a secret weapon for the Estonian state as it sought to attract foreign direct investment to rebuild the economy and create jobs.

The privatization of poorly managed Soviet operated enterprises was sped up, as were attempts to find foreign partners with the necessary capital and expertise to modernize the firms. Critically, unlike other newly independent nations, rather than handing the keys of the failing enterprises over to factory and business managers from the Soviet era, when the management of an enterprise was as much a political endeavor as it was a business venture, the government prioritized foreign partners who would also invest in new technologies, training, and processes. According to Linnar Viik, who served as one of Prime Minister Laar's key advisors,

> When foreign investments came to Estonia, small companies in Estonia also started using computers. This was very important because the foreign investments that reached Latvia had rather fax-based reports, and the foreign investments that reached Lithuania had rather checkbook-based reports.[28]

The local banking sector was particularly affected, jumping quickly over "technologies" like the use of checks which would amaze Estonians when visiting the West in the 1990s as they had never seen them before.[29]

President Armen Sarkissian, one of the founding fathers of modern Armenia, documented Estonia's rise in his book, *The Small States Club: How Small Smart Powers Can Save the World*. According to Sarkissian,

> What made Estonia stand apart was a combination of factors: the political will of its first leaders, the belief reposed in the state's ability by society, the country's discipline and its willingness to face as a united nation hardships on the path to an affluent

European future, and a competent financial policy as well as commitment to innovation, which optimized the management of limited human capital.[30]

Undoubtedly, the implementation of the Kroon and such difficult reforms were emblematic of the political will and discipline that President Sarkissian identified as some of the key reasons for the success of the newfound Estonian nation.

Critical for the new country's mostly empty coffers, in 1994 a flat-rate personal income tax was introduced.[31] The flat-rate tax was meant to make administration clear for citizens and the state, driving down tax fraud and making it easier for citizens unfamiliar with paying taxes to adjust to the new system, with the added benefit of saving them time at tax season. This proposal was panned by the IMF and, according to Laar in the 2018 documentary *Rodeo: Taming a Wild Country*, "When I came out with my idea to institute the flat tax, the IMF mission looked at me as if I was mad."[32] It would not be long before they were once again applauding Laar's foresight.

The flat-rate tax helped lead to a dramatic rise in tax compliance, driving down tax avoidance and helping eliminate the gray economy[33] that often plagues lower- and middle-income countries. It also led to an increase in state revenues as the country was able to collect more from citizens due to simple tax administration procedures and minimal bureaucracy. Laar would later say that,

> As the Estonian people saw that if they worked more, they would earn more and would not be punished by the government through higher taxes, their attitude changed surprisingly quickly. Thousands and thousands of new small- and medium-size enterprises, restaurants, hotels and shops were established. In 1992, Estonia had in total about 2,000 enterprises. By the end of 1994 the figure was 70,000.[34]

The introduction of the flat tax also made it easier for the government to execute what former Prime Minister Kaja Kallas later called a "complete turnaround in mindset."[35] Average citizens now saw the importance of not trying to steal from the state, which had been a noble endeavor when it was run by occupiers but if continued could destroy the fledgling nation. Paying taxes became a notable part of that population-level mindset shift after independence and was made easier with low and simple-to-understand taxes administered with little bureaucracy.[36]

Although flat taxes are often panned as regressive for penalizing the poor while essentially providing a handout to the rich, in Estonia's early days post-independence the policy was clearly a net positive, not just because it helped to eliminate the gray economy and boost desperately needed revenue for the state while increasing local levels of entrepreneurship, but because it created a new global narrative about the country and made the economy a significantly more attractive destination for foreign direct investment. Foreign direct investment and economic changes spurred by reforms have had a dramatic impact on the Estonian people's wealth and well-being. GDP rose from around $4.5 billion in 1995 to $14.11 billion in 2005 and $31.37 billion in 2020,[37] and average annual wages grew from the low single-digit thousands in the mid 1990s[38] to more than 20,000 euros annually by 2023[39] while life expectancy at birth rose from just under 65 years in 1991 to 74.5 years in 2019.[40]

However, policies such as the flat tax would eventually lead to increased wealth inequality in the country. According to a 2023 article, "Five per cent of the Estonian households hold a whopping 48 per cent—almost a half—of the country's wealth, while 95 per cent hold the rest between them," one of the higher rates of wealth inequality in the Eurozone.[41] Other economic reforms undertaken during the period have contributed to inequality. According to Toivo Raun's article "Estonia in the 1990s" in the *Journal of Baltic Studies*, "the actual process of economic change was painful for much of Estonia's population, and it substantially heightened inequality of income and wealth as well as social divisions."[42] The rise of inequality, especially between the capital, where roughly one-third of the country resides, and more rural areas, has since become a significant political (and sometimes security) issue as highlighted by then President Kersti Kaljulaid[43] who in 2020 "pointed out that while the level of prosperity in the Tallinn region is at 135 per cent of the EU average, the rest of Estonia has reached only 55 per cent."[44]

REBOOTING A NATION

Future concerns about inequality aside, Laar's immediate focus in the mid 1990s was on making the economy as attractive as possible for foreign firms in a bid to alleviate rampant poverty and joblessness. This emphasis on making the economy attractive for foreign direct investment was one shared by his fellow free marketers and politicians like Margaret Thatcher, who Laar greatly admired. While many of the post-Soviet economies depended heavily on aid to escape their dismal economic conditions, the Laar government made a conscious effort not to get hooked on aid—in part out of a fear of becoming dependent on others, in part for ideological reasons. For Laar,

> Foreign investment is very important for transition countries in such a situation. It is much more important than loans and development aid, which run the risk at some point of becoming factors that actually help to maintain the relative backwardness of the given country. Development aid may consist of obsolete technology and obsolete advice which is no longer needed in modern countries. But by using this assistance, countries in transition lose the opportunity to use their backwardness as a springboard for development.[45]

This mentality soon became something of a meme in the country, with Laar frequently proselytizing as to the benefit of an approach that prioritized trade over aid.

The picture is significantly more complicated than many anecdotal discussions about Estonia's use of aid (or lack thereof) suggest. While Estonia has intentionally built a global image of a trailblazing country that has forged its own path despite all odds and with little outside help, this ignores the fact that the nation has benefited a great deal from external support, even while still being nominally controlled by the Soviet Union. According to a 1995 report from the U.S. Library of Congress, "Estonia's transformation to a market economy during 1991–93 was eased considerably by the availability of more than US$285 million in foreign aid, loans, and credits."[46] According to Neil Taylor, author of *Estonia: A Modern History*, in late 1991, then President of Finland Mauno Koivisto said that "it would never be publicly known how much Finland did for Estonia during perestroika."[47] After Koivisto's death more than twenty years later, it was revealed that around 15 million Euros had been spent by Finland "nominally on Estonian culture, but in fact in many other fields as well to prepare the country for political and economic independence."[48]

As of 2023, the EU had provided more than 6.8 billion euros in funding to the country in structural aid over the past fifteen years,[49] which helped enable Estonia's development of its world-leading e-government services and rapid progress in building up the economy. Nonprofit organizations like the Open Society Foundation have delivered support at critical moments in Estonian history, such as supporting the Tiger Leap campaign and the development of the e-Governance Academy in Tallinn.[50] This is not to belittle any of Estonia's myriad accomplishments: the country overcame tremendous obstacles to get where it is today. But the historical record shows that it never solely depended on trade to advance, instead adopting a free-market mentality that penetrated both the bureaucracy and local zeitgeist, while also wisely using foreign aid when it was expected to be beneficial and in line with the country's goals, rather than blindly accepting all handouts offered and becoming dependent. In addition, in the earliest days of the nation, access to external funds was extremely limited and much of the funding, especially for technology-related initiatives, came from internal budget allocations, which spurred creative thinking since such budgets were extremely small.

Nevertheless, the idea of "trade, not aid" was hugely influential throughout Estonia's modernization and form a core aspect of the bureaucratic and business culture. According to Jennifer Pahlka, former U.S. Deputy Chief Technology Officer during the Obama administration, and author of *Recoding America: Why Government Is Failing in the Digital Age and How We Can Do Better*,

> In the business world, they say that culture eats strategy for breakfast—meaning that the people implementing the strategy, and the skills, attitudes, and assumptions they bring to it, will make more difference than even the most brilliant plan. In government, culture eats policy.[51]

Aligning on a shared direction and forming a distinct culture gave political leaders for decades the mandate they needed to continue investing in the creation of an Estonian state built for the modern world based on an Estonian vision of the future.

Laar was not the only one with the goal of building a country that served the needs of its people. Former President Lennart Meri,[52] or "Mr. Estonia" to the many diplomats and businesspeople who he met abroad while serving as the country's preeminent diplomat and evangelist after re-dependence, was also greatly influenced by seeing how foreign state services (and businesses) catered to the needs of citizens instead of the other way around. Priit Vesilind was an Estonian–American writer and photojournalist at the *National Geographic* for three decades who covered the fall of the Berlin Wall years after fleeing the Soviet army as a child, even taking up a sledgehammer to bring down part of the wall himself,[53] and knew Meri well. In a contribution to a book about Meri's life, Priit remembered how the President managed to finagle a rental car for a Soviet-endorsed visit to the United States while Estonia was still under occupation. Meri called him to tell him that when the car broke down the rental company immediately sent another car out to replace the damaged vehicle. According to Vesilind,

> I was waiting for the rest of the story, but that was all. He was excited because he could see that here, in the U.S., the nation's enterprises and leadership actually had a working compact for their mutual benefit. They were partners. He so much wanted that for Estonia. In the Soviet Union of that time the government had abrogated their part of that responsibility. There was no expectation from the people that any higher institutions would operate with a consumer or a client in mind.[54]

For Laar, Meri, and scores of other politicians and businesspeople, adopting a similar mentality was critical for Estonia's eventual success. By having the government focus on delivering useful products and services for the good of the people (and the business community), a virtuous cycle would be created that would make the country more efficient, wealthier, and better off overall. This mentality is one that has not been forgotten and has been continually adapted for the modern era. Laar's attitude of trade-not-aid permeated the bureaucracy, and the need to build a society where both business

A NATION REFORMED

and government operate as partners for the benefit of the people is still a commonly held belief.

Despite the high-minded nature of the ideals being adopted, they were often manifested in quite humorous ways. In 1997, Lennart Meri was serving as President. In Estonia, the role of President is often considered to be more ceremonial and less powerful than that of the Prime Minister (although Meri was known to personally wield a great deal of power)—more akin to that of the Vice President in the United States. A core responsibility of the Presidency is to promote economic and diplomatic ties across the world. While visiting Japan along with a delegation of businesspeople, Meri bluntly addressed potential local investors and business partners, asking what the country could do to make it more appealing to them. One of the businessmen who had visited several times blurted out that a good place to start would be the toilets in the Tallinn airport, which were in a terrible state after years of Soviet neglect.

According to later reporting, Meri was instantly red in the face and vowed to personally inspect the bathroom.[55] After his return to Tallinn, Meri held a now infamous press conference inside the bathroom in question. Bringing foreign journalists in to see the bathroom, Meri emphasized just how far the country had come, while clearly underscoring how much there was left to do.[56] Today, the airport has been renamed to the Lennart Meri Tallinn Airport in honor of his service and is a regional gem sporting amenities that would be more at home in the offices of a major tech company and run with incredible efficiency.

With the benefit of several decades of hindsight, it's apparent that the country was thriving thanks to the policies and investments made after independence. But even in the early 1990s, it was clear that the strategy undertaken by Laar, Meri, and cadres of other farsighted officials was working. By reforming the tax code, introducing their own currency and monetary policy, and liberalizing trade, the country had become a global growth success story with wages, employment, and GDP all on the rise. Professor Toivo Raun's research on Estonian life in the 1990s makes clear how these policies contributed to the growth of the nation as,

In both 1994 and 1998, for example, Estonia ranked first among the post-communist states of Central and Eastern Europe with regard to direct foreign investment per capita and as a percentage of GDP, thanks in large part to its liberal economic policies.[57]

These enduring economic policies also won the country a host of global admirers, including many of the think tanks who had advised Laar. The country seems to perpetually be at or near the top of rankings like the Freedom House Index on Internet Freedom, the Tax Competitiveness Index, and the Heritage Foundation's Index of Economic Freedom and has maintained its leadership for many years.

This is not to say that the reforms were easy. The average Estonian's life was hugely disrupted in the early years after independence and for a period many were worse off economically than they had been under the Soviet Union. The Soviet economy was inherently unsustainable and inefficient, so when Estonia opened its market there was a significant shock as industries ranging from manufacturing to agriculture were shown to be globally uncompetitive. This was little surprise when it came to agriculture: forced collectivization had clearly decimated a previously flourishing agriculture industry which had accounted for nearly 60% of Estonia's industry before World War II thanks to aggressive land reform in 1919 after the country gained independence for the first time, which would last until the onset of World War II. The reforms expropriated feudal manors and lands within the boundary of the country which were technically owned by foreign nations from prior conquests,[58] nationalizing much of it and distributing land to tens of thousands of Estonian settlers.

With the liberalization of trade and the introduction of the Kroon, many goods that had been essentially subsidized under the Soviet system became dramatically more expensive. For Estonian politicians enacting economic reform policies, it was a dangerous game politically. Because of the pain involved in issues like removing subsidies, which while necessary and often rhetorically supported were often unpopular in practice, doing what was best for the economy long-term was an excellent way to ensure you would not be re-elected in the short term.

A NATION REFORMED

The danger of a poorly executed reform policy dragged out over many years, causing continuous economic pain without clear benefits, led Laar to engage in economic "shock therapy." In doing so, he reformed as much as he could in one fell swoop in the hope that it would make the reforms stick, despite the pain they caused, by ripping off the Band-Aid all at once.[59] This had the added benefit of making it reasonably difficult for any future government to turn back his already enacted reforms, solidifying Laar's policies. He would later acknowledge that, "If people had known what was coming, they would not have elected us."[60] Laar and other young Estonian technocrats approached difficult reforms with intense (and perhaps irrational) confidence, even in the face of opposition, including from important global players like international financial institutions and many foreign advisors who were older and supposedly wiser than the young reformers. His attitude was similar to that of the founder of modern Singapore, Prime Minister Lee Kuan Yew, who recounted in his book *From Third World to First* that, "If I have to choose one word to explain why Singapore succeeded, it is *confidence*. This was what made foreign investors site their factories and refineries here."[61]

With the instatement of many pro-market reforms, the introduction and success of the new currency, ongoing privatization of many of the inefficient formerly state-run companies, and a burgeoning entrepreneurial sector, Prime Minister Laar and his colleagues were no doubt pleased with their successes and feeling confident in the path they had chosen. They had acted boldly and created opportunities for people, two of Laar's principal goals, and had put in place reforms that would far outlast the Laar administration.[62] The economy was thriving (or as much as could be expected despite economic headwinds like the flailing Russian economy next door) with business and job creation on the rise and much of the world had become cheerleaders of their work and held up Estonia as an exemplar of what liberal trade policies and a western-style economic system could do for other newly independent nations.

But while the economic situation was improving, many Estonian state institutions and infrastructure were still in disrepair and there was a constant danger of backsliding. One of the greatest risks came from inside the bureaucracy which was still replete with Soviet-era holdovers. While street violence, and petty and organized crime had been dramatically reduced, there was still a risk of corruption becoming endemic in the new system which would stymie economic growth and destroy Estonia's burgeoning reputation as a great place to do business. Political leaders desperately needed a way to both defeat corruption and increase state capacity, each of which would be a difficult task independently. Thankfully, the youth of Laar's cabinet and the Estonian political elite worked in the country's favor as political leaders embraced the potential of new technologies to solve the country's most pressing problems. After all, as former President Toomas Hendrik Ilves[63] is fond of saying, "you can't bribe a computer."

5

PRACTICING E-JUDO

The problems Estonia faced in the years following independence were both numerous and complex. There were few political leaders because most that the Soviets considered a threat had been exiled or killed during decades of occupation. After fifty years of a planned economy, there was little local entrepreneurial expertise or understanding of what it took to grow an economy. The country had seen decades of underinvestment in any physical infrastructure that was not useful for military purposes and the manufacturing and agricultural base was woefully inefficient. There was a significant cultural divide between native Estonians and Russian-speaking Soviet emigres that was a constant threat to domestic (and occasionally international) relations. The country had no major natural resources that could provide a lifeline while undergoing reforms, nor did it have a population large enough to be a significant geopolitical player based on size alone. With few natural advantages to build from, the Estonian people would have to use every tool at their disposal to prevent the nation's quality of life from become even more impoverished than during the Soviet era.

Judo is a highly cerebral martial art infamously practiced by Russian dictator Vladimir Putin, whose obsession with the sport runs so deep that he authored a book on the topic titled *Judo:*

History, Theory, Practice.[1] Judo is characterized by a mentality which instructs practitioners to identify and leverage weaknesses and turn them into strengths, allowing a traditionally outmatched contestant to defeat one who is larger and stronger. While Putin has attempted to apply the principles of judo to running the Russian nation, it is Estonian leaders like Prime Minister Laar and Presidents Ilves and Meri who have actually used judo's principles effectively. By directly identifying and tackling Estonia's shortcomings with policies that would enable the development of unique resources and capabilities that few other nations could emulate, Estonia's weaknesses became its strengths. One of former President Meri's famous witticisms is the unforgettable line, "The situation may be shit, but it's our fertiliser for the future,"[2] which encapsulates the Estonian judo mentality of turning a negative to a positive.

It is common for Estonian government officials in recent years to talk about the fact that as a country with a small poverty-stricken population spread over a large landmass, betting on digitalization was an obvious, or at least logical, choice, especially for those who understood the potential of the nascent web.[3] Former President Kaljulaid mentioned it often in her public remarks while she was in office and former Prime Minister Kallas has remarked that "digitalization became a powerful tool to rebuild our economy and society," when dealing with a lack of resources.[4] However, it is a path that few other nations took and one which had no certainty of success. So, while digitalizing government services and investing in developing local tech capacity and literacy may have been obvious for Estonia (at least in retrospect) there were unique historical factors behind the development and success of Estonia's e-government and tech sector.

One of the most important of these factors is cultural. In Estonia, there is a latent and widespread animus at having been lumped in with other Baltic countries and Eastern Europe rather than being considered "Nordic." Many Estonian leaders, including former President Ilves, feel they have little in common with the Baltic nations of Latvia and Lithuania other than a history of "occupations, deportations, annexation, Sovietization, collectivization, russification,"[5] all of which were imposed on the Baltic peoples by foreign actors. This is clearly displayed by the public remarks of various government officials across administrations,[6] as nearly all are eager to use any opportunity to point out how much more culturally similar Estonia is to Finland and other Nordic countries. At the same time, many Estonians believed that the country should naturally have a quality of life on a par or surpassing their Nordic counterparts based on their comparable living standard to Finland prior to the Soviet occupation.

This meant many Estonians thought they were destined to have a standard of living above "Eastern European" nations, leading to a perspective that leaders must do whatever it takes to try to move towards a Nordic or at least Western quality of life as soon as possible. This gave Estonian politicians more leeway in enacting bold reforms as the population was willing to tolerate more pain and risk to get back to their deserved place among the Nordics. However, it was obvious that following more traditional pathways for economic reform would never be enough to catch up to the Nordics, especially given how much of a lead they had after decades of communist repression in Estonia.

According to Laar, "Considering the chasm dividing Estonia from normal life, there was no other option than to try to jump over it, because it is impossible to cross a chasm in two steps. What Estonia needed was one decisive leap."[7] For the young government officials in the newly independent nation, technology was seen as the ideal way to cross the chasm and leapfrog to a Nordic standard of living. Bucking foreign expert advice on development and betting on a different path than nearly every peer nation is a major gamble, but as Josh Wolfe, noted investor in multi-billion-dollar tech companies like Hugging Face and Anduril, often remarks, "chips on shoulders puts chips in pockets."[8] In other words, people with something to prove are often ones that have a chance to really create the future. The Estonian people had a major chip on their collective shoulders, and they were willing to make bold bets and support leaders who would do the same.

There was another factor at play that was closely tied to the mentality pervasive in the political elite in newly independent Estonia—youth. Estonia's political elite were usually in their twenties and thirties, coming of age at the dawn of the modern web. Toomas Hendrik Ilves, who would serve as President, Ambassador, and Minister of Foreign Affairs across a multi-decade career in government leadership starting in the early 1990s, had first encountered programming as a thirteen-year-old student in New Jersey.[9] He began his involvement in the Estonian independence movement in his youth as a journalist at Radio Free Europe,[10] well before his formal roles in government. Mart Laar was only thirty-two and already the "old man" in the group when he was elected Prime Minister, with many ministers in his cabinet being even younger.[11] Although Lennart Meri was not a young man by Estonian standards when he took up a role in the newly independent government, he had been tinkering with electronics for decades and understood the potential of technology as well as someone who had often been able to glimpse beyond the Iron Curtain by gaining surreptitious access to broadcasts from groups like Voice of America.[12] During a visit to the United States, he would say that meeting with Bill Gates would be more important than meeting with President Clinton given how important he perceived the impact of technology to be for Estonia.[13]

Because of their youth and the fact that they were familiar with technology, leaders like Laar and Ilves were willing to make investments that would have been incomprehensible to an older generation who hadn't seen or couldn't understand the power of the World Wide Web and the nascent internet economy. From investing in e-government platforms to getting schools connected to the internet and teaching basic digital literacy skills, the youth movement leading the nation laid the foundation for Estonia's eventual primacy as an e-government and tech leader. According to Skype co-founder Jaan Tallinn, "It helped that many politicians in the early 90s were unusually quick to 'get' the internet," and that "The people in power after the collapse of the Soviet Union were really young."[14] Linnar Viik, the government advisor and architect of many of the Estonian investments in e-government, has also highlighted the importance of sustained political will when it came to technology, saying in an interview with *The Guardian* that, "among politicians of all stripes there has always been 'a silent consensus' about the importance of the internet. For many years, Estonians could expect whomever they elected to have the best interests of the internet at heart."[15] While the lack of older political figures with deep expertise was likely considered a danger to a fledgling democracy by many, Estonians masterfully turned a negative to a positive.

However, the dearth of trusted public servants with relevant expertise presented not just a power vacuum but a capacity gap which the new state administrators had to rapidly fill. Without the ability to deliver government services the Estonian people would surely rebel, possibly against democracy and almost certainly against newly elected politicians. Part of the answer would come from the private sector, namely banks, which became a frequent partner throughout the nation's development, especially when it came to digital matters.

In Estonia, there has long been little separation between the public and private sectors, with frequent moves between them common for workers of all seniority and experience levels, but the financial services industry has played an especially storied role in Estonia's development. Internet banking was novel in the mid 1990s, and in newly developing Estonia, engineers excited about the possibilities of the internet developed online banking services, sometimes in their free time, and pushed management to adopt them.[16] Soon, the country was a world leader in online banking, with Professor Tarmo Kalvet commenting,

> It is somewhat extraordinary how quickly electronic banking and Internet banking has emerged in Estonia... It is even more outstanding that as the world's first Internet banking services started in 1995, and by the end of 1996 there were only about 20 such services, of which three were from Estonia.[17]

The capacity developed in the financial services sector in digital and cybersecurity would often come in handy for the Estonian state and it was not unusual for individuals to volunteer when their services were needed by the government (especially in extreme cases like large-scale cyberattacks against public digital infrastructure).

In an interesting turn for a relatively insular culture, the country also looked outward and welcomed a variety of foreign experts to embed within the government, providing critical expertise in a variety of institutions, a practice that remains in place even today.[18] But the main method for solving the problem would once again be technology, which was held up as an ideal solution. Toomas Hendrik Ilves had seen the potential of computing to dramatically increase productivity and believed it would be a critical tool for a nation with such a small population, rather than a dangerous path to social or economic disruption as many larger countries viewed it. He and many of the other politicians in the first years after Estonian independence understood that by investing in e-government solutions they could tackle three problems at once: a lack of experienced bureaucrats, public sector corruption, and delivering services more efficiently.

Estonia fully embraced the use of e-government technologies, making the digitalization of essential services and transition to an "information society" a core part of the government's strategy from nearly the beginning of the nation's founding, with the building of e-government services taking off in earnest in the mid- to late 1990s. But, as was becoming a pattern with the leaders of the Estonian nation, they did it in their own way. In 1993, a memo from an Estonian professor of computer engineering, Raimund Ubar, made it to Prime Minister Laar's desk, leading to a foundational policy for both the Laar and future administrations: a commitment to not buy legacy equipment. According to Professor Rainer Kattel and Ines Mergel's paper "Estonia's Digital Transformation: Mission Mystique and the Hiding Hand,"

> While Ubar's memo focused on industry and applied research, what seemed to have impressed Laar was Ubar's insistence on avoiding the legacy trap he had observed in Western countries and encouraging a policy of investment in future-oriented emerging technologies rather than in legacy technology. Laar recalls: "Ubar wrote that it is not sensible to overreact and start immediately by buying things. He was the first person who told us not to buy anything old."[19]

Laar and Ilves were convinced early on that it was dangerous to depend on legacy IT equipment and software, which could eventually cost far more than it saved. That meant that Estonia could buy new or build it themselves. Since the country was broke, there was truly only one option—embracing a culture of innovation and frugality, building whatever they could internally. According to Sten Tamkivi, a noted Estonian entrepreneur, investor, and early Skype employee who has been involved in government strategy since he was a teenager,[20] this turned out to be a huge boon for Estonia in the long term. Recalling Estonia's development, Sten commented that,

> The other factor that I think we made really good use of was the lack of resources. Which again sounds like a deficiency at first, but it's one thing that makes you think creatively. If you're a western government ministry and you need to build a bookkeeping system, then IBM and Oracle and these guys come to you and say it's going to cost twenty-million dollars... In Estonia, sorry, haven't got that so let's put two guys in a room and see what they come up with.[21]

Venture capitalist and early Skype investor Steve Jurvetson has also remarked that Skype was a beneficiary of this resource deprivation, which forced the team to be more innovative and resourceful compared to better-capitalized tech firms like Microsoft (who would eventually acquire Skype after it spun out from eBay), highlighting that resource constraints leading to innovation applied just as well to startups as it did to government.[22]

Another advantage of digitalization was its power to curb the public sector corruption that threatened Estonia's development. Where there was a threat of corruption destroying the fabric of the state before it was able to truly develop and preventing the achievement of major policy goals like accession to the European Union and NATO,[23] e-government systems played a critical role in preventing low-level public sector corruption. According to the Tallinn-based e-Governance Academy, which was founded as a joint initiative of the Estonian government, the Open Society Institute, and the United Nations Development Programme,[24]

> Since everyday transactions with the state are mediated by computer systems, bureaucrats no longer have as much opportunity to take small gifts of appreciation to accelerate the process. Hence it can be said that digital transformation enforces the rule of law.[25]

Today, Estonia boasts one of the lowest levels of corruption in the world, ranking fourteenth in Transparency International's Corruption Perception Index alongside Canada and Iceland.[26]

Even in dramatically different contexts, e-government platforms have been used successfully to tackle corruption, highlighting the usefulness of such systems. For example, in 2015 Ukraine launched the open-source Prozorro procurement platform, which was called "the gold standard" by the former Chief Technology Officer of France.[27] Within just the first fourteen months of operations, the platform processed more than 100,000 tenders and has helped to reduce waste while increasing transparency leading to savings of billions of Hryvnia[28] and played a key role in President Zelensky's agenda to "transform Ukraine into a 'digital state in a smartphone,'" which was undertaken as part of a drive to "enhance transparency, combat corruption, and simplify complex administrative processes, all with 'one click.'"[29] That is not to say any e-government solution is a panacea; many countries which have implemented such solutions still struggle with corruption and myriad other issues, but such solutions can be a valuable tool in any government's toolbox.

Looking at how Estonia achieved its rapid evolution to an e-government and startup leader in thirty years, policymakers in other countries may be inclined to believe that there is something intrinsic to Estonia or the Estonian people that has enabled such rapid progress. Part of the answer is surely the exceptional efforts of politicians taking bold action and scores of experts and former private sector executives who wanted to do their civic duty and give back to their country and the unique culture they inculcated within the bureaucracy and who worked tremendously hard to achieve what they have.

However, it was more than just hard work that paved the way for Estonia's global tech leadership. They also achieved independence just as the digital economy was taking shape, which played a critical role. According to Ott Kaukver, former Skype employee and ex-CTO of Checkout.com,

> I think we were lucky with timing. I think if we would have gotten independence back 15 years earlier, I think the IT revolution or

like the way that we use technology would have been a little different than it is today... You need to run the country and you don't have any resources... You don't have any mainframes or any systems that, like in some countries, still exist nowadays. You can select the best technology and use this in a very creative way. That was an opportunity for the country to install the best technology at that time and really be efficient with that. And if you don't have any legacy you can move fast.[30]

Kaukver's sentiment is one often echoed by Estonians engaged in developing e-services or who were involved in Estonia's e-government and tech strategy. Estonia bet big on the web just as the digital economy was truly taking off and didn't have the burden of existing legacy software products. Getting the timing right for when to invest in certain technologies or reforms is a crucial lesson for other countries as they invest in their digitalization journeys.

For countries hoping to emulate Estonia's success, creating a carbon copy of Estonia's existing infrastructure based on how the country developed is a fool's errand. Instead, they should look towards current and future trends in computing and consumer behavior while designing an e-state. For those beginning their digital government journey in 2024, investing in AI and mobile technology seems significantly more promising than rules-based software optimized for the desktop. The world has changed since Estonia began its digitalization journey, along with consumer preferences, even for government services. This has been especially apparent in the rapid move to mobile devices as the primary interface to the digital world for consumers, something which Estonia was slow to catch on to as it had invested heavily on building applications most easily accessible from a desktop.

In a single generation, Estonia has achieved what many believed impossible, or would not even have dared to imagine. While the country started off with few natural resources, it masterfully leveraged the few resources it did have to succeed and become a digital government and tech ecosystem leader. Where politicians lacked experience, they made up for it with youthful exuberance and a willingness to be bold and move quickly. Where corruption threatened to slow down economic growth, new technology was used to fight it. Where the country lacked a large population which could give it geopolitical and economic power, it used speed and leveraged technology to punch above its weight. And while there was some luck involved in being in the right place at the right time at the beginning of the dawn of the digital age, it took skill to identify and take advantage of the opportunity. In truth, the country created the only natural resource that mattered: a people who had the knowledge, the tools, and the future-thinking mentality to build a nation equipped for the information age.

PART I

LESSONS LEARNED

Just Do It, Quickly: Prime Minister Mart Laar was fond of saying "Just do it" in reference to Nike's universally known slogan when speaking about engaging in reform. However, possibly more importantly than "just doing it" is to do it fast. The reforms his administration undertook were both necessary and incredibly difficult. Undoing the damage of decades of poor Soviet policies was only possible by ripping the Band-Aid off all at once—slowly engaging in reform only meant that the pain would be drawn out over years or decades. This is also critical for leaders wishing to remain in power—sometimes moving fast is the only politically expedient way to create change. When a country is ready for reform, just do it. But do it quickly.

Culture Eats Strategy for Breakfast: Steering the direction of a new nation is like steering the direction of a startup. Factors on the ground are changing rapidly and there is much to be done. More important than a single brilliant strategy that will be out of date the moment it's published is to build a high-performing culture that is resilient and focused on delivering for the good of the people, no matter how difficult the task at hand.

Constraints Spur Innovation: Whether for a startup like Skype or a country like Estonia (or Singapore or Israel), facing massive resource constraints helped lead to innovative ways of thinking and spurred action that larger, better-resourced 'competitors' would have been loath to take. The judo mentality means turning every negative to a positive, and innovating around resource constraints is a constant in successful companies and countries.

Trade ~~Not~~ and Aid: Building a culture and society that embraces trade and economic dynamism is critical to the long-term success of a nation. Aid can be a godsend and, in some instances, necessary for the ability of a nation to survive, but depending on aid can be dangerous—donors get impatient, funding dries up, the local culture shifts from building to begging, and bureaucracies become incentivized to view donors as their constituency rather than the local populace. Aid should be used as a supplement or an unexpected gift that can free up resources for higher-leverage investments, but it can never be part of a serious long-term national strategy and certainly never relied on to last for any amount of time.

Timing Is (Almost) Everything: Most Estonians freely admit that part of the reason why the country has become a leader in e-government services is because it gained independence and started building government services right as the web was taking off. But Estonia also missed the move to mobile—as consumers were switching to smartphones from desktops, e-services were still being built and optimized for computers. Later entrants to e-services like Ukraine have been quick to focus on mobile due to its now overwhelming popularity, but governments should recognize when building services that things change—the next generation of applications may be powered by artificial intelligence or take place on a yet to be invented device or interface. It is important to build products, services, and bureaucracies that are both resilient and future-proof.

PART II

BUILDING A DIGITAL SOCIETY

6

THE BEDROCK OF THE E-STATE

Estonia's e-state, comprising more than a thousand public and private services accessible via government digital infrastructure,[1] has received nearly universal positive acclaim in the international press. This ranges from *The New Yorker*'s effusive praise in articles with titles like "Estonia, the Digital Republic"[2] and "Why Estonia Was Poised to Handle How a Pandemic Would Change Everything,"[3] to *Wired*'s piece "Welcome to E-stonia, the World's Most Digitally Advanced Society."[4] The e-services making up Estonia's e-state (also interchangeably referred to as e-government or digital government) were built on a bedrock made up of two parts. First, a unique set of formal and informal norms and principles guiding e-government services has developed over the last several decades since independence, often organically and sometimes haphazardly. Second, two key technologies, which most e-services depend upon to function, were created or built upon for use in Estonia, digital identity (sometimes referred to as e-ID) and the X-Road, a secure data exchange layer.

Informal norms and policies first took shape in the minds of a few key leaders who guided the development of the Estonian e-state from its earliest days, including former Prime Minister Mart Laar, former President Toomas Hendrik Ilves, and Laar's advisor, Linnar Viik. The first time that pen was put to paper on an official strategy document was with the 1998 "Principles of Estonian Information Policy." More high-level manifesto than detailed plan, the document numbers only a few pages. Key policy objectives are outlined including the overarching goal to, "help to create a society and a state that serves citizens, promotes their participation and cares for their well being."[5]

The policy plan also laid out subgoals such as promoting democratic values and the preservation of Estonian culture and language, the development of a competitive economy, and improving defense through information technology. These subgoals often number only a few words apiece and sections of the document pertaining to critical issues like taxation and legislation consist of only a few bullet points. However, that high-level policy document provided a lodestar for those actually developing e-government systems, especially after its official ratification by parliament, creating an outline with enough white space that government intrapreneurs[6] could flexibly navigate while creating new e-government technologies and policies. As former President Ilves is known to remark, "Laws are the software of society" and "If you want to change society, you have to change some of that software, too."[7]

The "Principles of Estonian Information Policy" marked the first formal creation of "software as law," targeting a critical constituency of Estonian society—those developing e-government systems. It would be an important part of the country's strategy of becoming a welcoming regulatory environment with a strong rule of law and clear rules, especially for emerging technologies and private enterprises. As Ilves and other early pioneers of the development of the Estonian e-state are often quick to point out, "the key to success in technology and digital affairs is not the technology but rather policies, laws and regulations."[8] Having strong legal frameworks and a welcoming regulatory environment have directly contributed to Estonia's success in becoming an e-government leader, but have also been especially helpful in attracting foreign direct investment, with former Prime Minister Kallas saying in a 2023 *Bloomberg* interview that, "[the] first thing we understood back then was that what makes investors trust your country is the rule of law. So, building the laws and the legal system was very important in order to attract investments."[9]

THE BEDROCK OF THE E-STATE

The initial principles developed by e-government champions like Laar and Viik were frequently codified and expanded with new statutes and rulings by Estonian lawmakers, often at the behest of those building or planning to build e-government systems.[10] This provided both guidance and guardrails to those building systems, while creating the legal frameworks and clarity needed to execute the development of ambitious technologies and initiatives—from a mandatory digital identity for every resident of the country to the e-Residency program which enabled foreign citizens to access Estonia's e-government infrastructure from anywhere in the world. According to Professor Tarmo Kalvet, "Estonia has often been reported as a country with [a] favourable legislative environment towards ICT and the most important legislative acts have been approved without external pressure..."[11] The most critical of those foundational statutes were the Digital Signatures Act, Public Information Act, Personal Data Protection Act, and the Identity Documents Act.

In 1999, the Identity Documents Act passed, mandating that all Estonian citizens and foreign nationals living in Estonia had to receive a digital identity which assigned a unique identifier to each person. This was a bold decision and one that could have easily backfired, but according to many of the champions of e-services like former President Ilves, "You have to make it mandatory for it to be successful."[12] However, modern critics of digital identity programs would be quick to point out that such programs can quickly turn into a new lever for state control and surveillance, with significant privacy implications for individual users.

Interestingly, the technology behind digital identity solutions was not new. According to Laura Kask, who served as the legal architect of the world's first data embassy, Estonia actually based the idea on Finland's already introduced digital identity system,[13] one of many technologies the Estonian state has adapted or adopted more successfully from pioneers in the public and private sector. But where it was voluntary in Finland and had relatively minimal uptake, in Estonia, digital identity was made mandatory and thus a market for e-services was created by making a user base via fiat. This was critical, because while the technology had long existed, collective action problems had led to slow uptake in other countries. The legal change to make it mandatory served to highlight President Ilves' remark regarding laws being the software of society and that to change society, the software must be changed too.

While in most countries this would be difficult politically for any number of reasons, from concerns over enabling a state surveillance apparatus to privacy rights and putting individual data at risk if not properly safeguarded, Estonia was able to pull it off because of the tremendous amount of trust and goodwill that early political leaders had generated. This was thanks to their track record of successful reforms and because there remained a persistent mentality that political leaders should do whatever was necessary to help Estonia catch up to and ideally leapfrog its Nordic neighbors. Beyond goodwill and the capacity to execute such an ambitious agenda, the role of data protection legislation was also critical to making citizens feel that they had the law on their side.

The Digital Signatures Act was passed by parliament in 2000 and provided the legal basis for the main value proposition of digital identity, that an Estonian resident (and since 2014 an e-resident) could use their digital identity document to digitally sign a document in a way that was legally binding. This meant no more printing out digital forms, signing them, scanning the document, and then emailing it on. It also mandated that all public institutions accept digitally signed documents. Although this sounds simple enough, consider the fact that in the U.S., which long had no such comprehensive law or government developed technology to enable such a function, companies like Docusign have sprung up to make it easier for people to digitally sign documents. The fact that the firm had a market capitalization of just under $12 billion in May 2024[14] shows the value of being able to do something seemingly as basic as enabling legally binding digital signatures in a simple and convenient way.

The Public Information Act and Personal Data Protection Act guaranteed key rights regarding privacy and personal data to the users of the Estonian e-services. Given that having a digital identity is mandated, the acts cover practically every resident of the country. Passed in 1996, the Personal Data Protection Act guarantees "the right to [the] inviolability of private life."[15] In practice, it meant that there were clear rules around the use of certain types of personal data, grouped into the buckets of sensitive and non-sensitive, which had differing rules for their use and processing by government authorities. It also meant that the user rather than the government was the one who owned their data and empowered users to take steps to control their data. Critically, the law created a Data Protection Inspectorate which "was established as the defender of the related constitutional rights," according to e-government expert Raul Kaidro.[16]

The Public Information Act meant that anyone could ask the government for information and required the government to respond within five working days, also mandating that government agencies put extensive amounts of information on the public web where it would be available for all. This included details on company and land ownership that would be private by default in a regulatory environment like the U.S. or Britain. While there can be fees for more complex requests for non-public information that the state has to collate and deliver, and some exceptions for private data or information relevant to national security, the act created an environment of near total transparency. It also enabled something that Estonians like to call "little brother," where citizens have an active role in supervising the state.[17] While in other countries, government information can be accessed through lawsuits and Freedom of Information Act requests, these are often complex, expensive, and slow, whereas in Estonia, thanks to the digitalization of government services, such information is generally readily accessible.

While concerns about privacy over the government having access to so much personal data are often raised, especially by privacy-conscious foreign visitors to the country, former President Ilves has an alternate viewpoint, saying "I feel much more secure with a digital ID. If anyone goes into my files, they're flagged. Whereas if my files—which would exist anyway—were made of paper, no one would know who was looking at them."[18] Because of the Public Information Act, when someone sees that their data has been viewed, they can send a request to inquire as to why and under what authority the review was taken and a response is required, which serves as a strong deterrent against snooping. This is quite different to the status quo in places like the U.S. and Germany where many official government records are on paper and there is little accountability or visibility into who accesses these files across various agencies and for what purposes.

While there have been a few instances of malfeasance in Estonia, like a police officer trying to look up information on his bride-to-be,[19] the perpetrators were relatively easily caught thanks to the security and transparency mechanisms built into the system. However, such a system only works well in a democratic country with strong state capacity, rule of law, and minimal public sector corruption. In many other contexts, such an agglomeration of data in the hands of the government (or any individual actor) could potentially enable significant harms. For example, the ID system in Uganda has been called a "perfect system in the hands of an imperfect and brutal regime that applies the system to suppress opponents, target critics and settle personal scores."[20]

These legal acts were important not just for providing rights to the Estonian people but because they enabled the development of key services that now consist of the Estonian digital government. The most important of these services were the X-Road system and digital identity, which are widely considered to be the foundational technologies that enable the modern e-state. The X-Road solved a critical problem that all complex systems face, allowing for the secure transfer of data, including important personal data like health and financial records, between disparate databases controlled by various public and private sector entities. Although the sharing of information between multiple systems seems straightforward, in practice it can become quite complicated, especially for those who care about privacy and reducing bureaucracy. A deep-dive by *The New Yorker* into Estonia's e-state described the system well:

> Data aren't centrally held, thus reducing the chance of Equifax-level breaches. Instead, the government's data platform, X-Road, links individual servers through end-to-end encrypted pathways, letting information live locally. Your dentist's practice holds its own data; so does your high school and your bank. When a user

requests a piece of information, it is delivered like a boat crossing a canal via locks.[21]

Today, the Estonian-developed X-Road is not just open-sourced in an effort to promote transparency and global use, it is controlled by the Nordic Institute for Interoperability Solutions, an intergovernmental collaboration,[22] and more than twenty countries utilize some version of the open-sourced platform.[23]

If data is the treasure, digital identity is the key. A digital identity is "Issued at birth and good for life... a unique digital document which allows people to securely log on to private and public websites, removing the need for extra checks on their identity."[24] In essence, rather than a social security number written on a piece of paper, Estonians get a unique digital identifier which they can conveniently use in combination with a physical identity document, which is embedded with a chip so that it can be used as a login to digital systems. Alternatively, the identity document can be made entirely virtual via an app on a user's smartphone. The system is constructed with security and convenience for the end user at the heart of every design decision, using multiple keys for authentication and digital signatures, including built-in redundancies in case of the loss of a PIN, a cyberattack, or attempted identity theft. According to an *Atlantic* article authored by Estonian entrepreneur and investor Sten Tamkivi,

> To prevent this system from becoming obsolete in the future, the law did not lock in the technical nuances of digital signatures. In fact, implementation has been changing over time... Every person over 15 is required to have an ID card, and there are now over 1.2 million active cards. That's close to 100-percent penetration of the population.[25]

How the Estonian e-government system works in practice is best illustrated by comparing the pathways of a newborn Estonian and a newborn American whose families wish them to have access to government services (we'll ignore the Kafkaesque U.S. health insurance system and focus on government services). When an American baby is born, the parents must apply for a birth certificate and then a social security number, generally with two different agencies. Both are necessary for the parents and the child to interface with any number of government and even private sector services, from being able to claim them as a dependent at tax time to the child eventually being able to open a bank account. These critical documents meant to last a lifetime are then mailed to the new parents and safeguarded under lock and key, often literally in a safe deposit box because of their importance. It is quite possible that the new parents are eligible for various programs at the state, local, or federal level, or at least various tax incentives. However, they will have to apply for each benefit individually (if they are even aware of their existence), inputting the same information over and over in various unconnected government databases. There is little to no interoperability or communication between systems, everything is manual, and a significant burden is placed on new parents. According to a 2024 op-ed in The *Washington Post* about the burdens of paperwork for new parents in the U.S.,

> There are many parts of parenting for which it's impossible to prepare, be it the first late-night trip to urgent care with a miserable, feverish toddler or a big question about sex or death asked at an inopportune moment. But perhaps the most mundanely irritating of these surprises is the vast amount of paperwork that follows children, like Pigpen's unrelenting cloud of dust. At minimum, this secretarial work levies a time and emotional tax on parents. At worst, paperwork can become an obstacle to getting financial help, medical insurance, aid for college or even an elementary school education.[26]

In Estonia, the process works differently. When the baby is born, it is issued a digital identity after details are uploaded by the hospital immediately after birth. Because the parents also have digital identities, the child is then automatically "attached" to their accounts. That's it. The X-Road is what enables the transfer of the information between the various parties like the health system, tax authority, and the hospital, allowing them all to transfer data seamlessly while protecting the privacy of the end user. Digital identities are what allow the government to understand what data belongs to who, and who should have access to it.

THE BEDROCK OF THE E-STATE

Still, the Estonian government has been working to make things even easier for new parents as part of a push to create proactive e-services that take as much of the pain out of the process for citizens and residents as possible, and ideally saving time for government officials, by automating various tasks. For example, after registering their newborn child, parents "are sent an email notifying them of their eligibility for additional family benefits. After clicking on the link and logging into the website, parents simply enter their bank information, and the benefits are automatically registered and transferred."[27] This has major cost and time-saving benefits for both parents and government, as previously every application for such services took state officials more than an hour to process.[28] A government proactively working to give citizens benefits is practically unheard of in most of the world, but in Estonia it is a core part of the country's strategy to become a mecca for innovative digital governance and is enabled by key pieces of legislation and the development of a unique culture across the country's bureaucracy.

This culture is user-centric and has been built over the last several decades, based on norms and principles developed by those building e-government systems. Some of the most important principles frequently utilized in Estonia are: Build vs. Buy and Embrace Frugality, Avoid Legacy Technology, Less State is More, Once-Only, Technology Neutrality, and Digital by Default. Many of these principles are interconnected and partially overlapping but all have been critical in the development of e-services that are built to make the lives of citizens easier rather than to check a box or fulfill an arcane legal requirement. Often, these principles are not codified into law or remain rather nebulous with various interpretations but are followed in their own way by the close-knit ecosystem of e-government experts and the political establishment in the country, having become ingrained in the culture.

One of the earliest principles was the focus on taking an extremely measured approach to building e-services versus buying them, especially from foreign sources. Intrinsic in this principle was the idea of embracing frugality—in no small part arrived at simply because there was no other option for a government which in the early 1990s was facing severe budgetary constraints and couldn't afford to procure newfangled systems from expensive western providers. Similarly, the principle of avoiding legacy technology, which while taken at face value might lead one to believe that the government would be more willing to spend money on new systems, has deepened the culture of developing the infrastructure needed to run the e-state domestically by pushing the government to consider software as an ongoing operating expense, rather than a single capital expenditure. Instead, investing in domestic capacity, whether directly in the government, through symbiotic partnerships with the private sector, or building on top of freely available open-source software, has historically been seen as an optimal strategy, leading to Estonia's government not only being able to field cutting edge e-services, but a domestic public and private sector ecosystem that has become a leading exporter of e-services and related consulting services across the world.

The Once-Only principle is a relatively straightforward concept to understand, but extraordinarily difficult for a bureaucracy to master. According to former Prime Minister Rõivas,[29] "In Estonia we decided that we should only ask for information from citizens once and not ask for that information again." Practically, that meant that if someone like Rõivas was to move within the country,

> I only need to tell the population registry—which logs everyone's addresses. Once they have been notified, the tax authority, or health service, or anything else, should not ask me where I live. As a citizen, it is not on me to remember which government agency needs to know my new address, and this is far more convenient.[30]

The Once-Only principle's value proposition for citizens is clear: no more filling in the same paperwork over and over, no more worrying (or perhaps hoping) that if you move, His Majesty's Revenue and Customs will no longer be able to reach you about your taxes. But for government agencies to accomplish this, it meant that their internal systems needed to effectively share data and to do so in a format that didn't cause tremendous amounts of manual work translating one agency's chosen format to another's—an ability enabled by Estonia's X-Road. According to

"Deploying the Once-Only Policy," a 2020 policy brief published by Harvard's Kennedy School, by effectively implementing the same principle and streamlining an overburdened bureaucracy, the EU could expect to save 5 billion euros annually,[31] and a 2021 Brookings paper referred to Estonia's implementation of the principle as a "gold standard."[32]

Making government services digital by default is another task that is easier said than done. According to the principle, any new service or interaction that an individual user undertakes with the government should be able to be done completely virtually.[33] This means that policymakers thinking of rolling out a new regulation must consider how it can be effectively managed and complied with online by the average person—something that is often neglected by those shaping legislative agendas in other countries. Making services digital by default also enables another of the prized goals for a lean bureaucracy—working to have "less state."[34] By empowering citizens to interface with government agencies online rather than in-person and by automating processes and eliminating bureaucracy whenever possible, the bureaucracy in Estonia is smaller and less costly than what would be required if all services were only delivered in analog form, with all the associated human capital required to do so at scale.

When it comes to the importance of being able to connect to the internet—whether to the global economy for work, for access to essential services like banking and healthcare, or to stay connected with others—calling for making internet access a human right was an obvious choice and something Estonia did in the early 2000s, well before many larger Western European nations.[35] And while access to the internet is now increasingly acknowledged as a human right around the world, Estonia has also focused on promoting internet freedom. According to the 2023 "Freedom on the Net" report, the nonprofit Freedom House's annual survey on internet freedom which measures factors like obstacles to access the internet, government-imposed limits on content, and violations of user rights, Estonia was ranked as the second-best environment for internet freedom after Iceland.[36]

Embracing Frugality, Digital by Default, Avoiding Legacy Technology—these and the rest of the aforementioned principles have become just as important as the foundational legislation codified into statute, from the Digital Signatures Act to the Personal Data Protection Act. Norms can be powerful, serving as both guides and constraints for those considering new e-government projects and legislation, eventually becoming ingrained in the local culture. Critically, the laws and principles discussed have become the basis for modern Estonia's e-government and enabled the development of the two technologies which underpin all of it, X-Road and digital identity (the latter of which was truly made successful thanks to the law making it mandatory, rather than the specific technology it was built upon). Taken together with the principles and norms forged over three decades, they form an operating system for how the Estonian government approaches the development of its e-services and digital society.

7

BECOMING E-ESTONIA

By 2002, more than a decade before the infamous botched rollout of the U.S. Healthcare.gov website (also known as Obamacare),[1] Estonia had already developed an e-state far more advanced than anything that exists in America or much of the Western world even today. A 2012 McKinsey report documented the breadth of the services made accessible by Estonia's digital government, saying "Today, customers can log into their bank accounts directly from the state portal. They can also claim loyalty rewards at the local cinema, purchase bus tickets, pay electricity and phone bills, and renew medical prescriptions."[2]

At the beginning of the new millennium, the Estonian government launched the first of three of the core e-services that make up the state's digital government. The first, a free e-tax solution that allowed citizens to declare and pay taxes online, increased much needed state revenue while minimizing the potential for corruption and fraud. Thanks to the digitalization of the service, it now only takes three to five minutes on average for Estonians to do their taxes.[3] Soon after, the government created an e-health system which enables doctors and patients to quickly access their medical information from a simple online portal, dramatically decreasing paperwork burdens for medical personnel and patients and increasing quality of life. And, probably most importantly for the highly democratic and egalitarian country, it developed an online voting solution which allowed citizens to vote from anywhere in the world (assuming they had internet access).

Leaders like Prime Minister Laar and President Ilves created the policies and environment for an e-state to flourish in Estonia, but government policy was not the beginning of the e-state; poverty was. More specifically, it was the intersection of poverty, geography, and a search for profit by the now capitalist banking sector that gave the e-state its start. In the early 1990s the financial services sector was in a sad state due to the broader economic situation in the country. However, given that the economy was also newly market-oriented, it was apparent that individuals and businesses would need banking services, attracting a host of domestic and international players. The banks, much like the Estonian government, encountered a problem while trying to build up their presence across the country—many towns and villages were simply too remote, too small, and too poor to merit the investment of creating (or in some cases maintaining) a physical branch. To provide access to banking services while remaining profitable, the banks had to look for alternatives to scaling up physical infrastructure.

Given the domestic proclivity for leveraging technology to solve complex challenges, it's no surprise that the banks decided to look to the internet as a potential solution to their problem. Moving to online banking meant that banks could more profitably reach customers across the nation without investing significant resources into physical infrastructure. But while online banking solved the problem of delivering services to the population at scale and at a reasonable cost, banks now faced a new twist on an old problem: theft—now in cyberspace.

Because the banks invested in allowing their customers to access services online, they were highly incentivized to make sure that the systems were secure. While banks in Estonia were pathfinders in developing online banking services, it also meant that they were vulnerable to potential cyberattacks and theft, especially given their proximity to Russia, which has long been a hub for cybercrime[4] and in 2024 topped the first ever World Cybercrime Index.[5] The potential for cyberattacks and fraud necessitated investment in two areas: cybersecurity and identity verification. Leading banks in the country strengthened their cyber workforces in an attempt to stave off attacks and create more resilient digital infrastructure, eventually creating a virtuous cycle of cyber experts moving between the banking sector and the government, each strengthening the other.[6]

At the same time, banks understood that basic online identity verification tools were insufficient when dealing with potentially large sums of money. Looking for other solutions, the banks eventually developed a robust digital identification system that enabled users to securely access their accounts and engage in online transactions.[7] The banking industries' digital identification system was considered extremely secure and with its rapid adoption in Estonia, a highly successful implementation of critical digital infrastructure. Soon, the government saw the success of the bank's digital identity solution as an opportunity to leverage an existing, successful methodology for their own purposes (namely as an alternative login credential) as they built up government-run e-services.

To integrate the solution into the larger e-state service toolbox, the government partnered with a consortium of leading banks and telecom companies to develop standards and a solution based on the private sector's best practices that could also be leveraged to access e-government services.[8] This would make life dramatically simpler for everyday users who already had a digital identity they knew how to use via their bank and could continue utilizing it as a convenient alternative (or in addition to) the government's digital identity solution.

Thus, online banking became one of the kernels of the e-state, serving as a proof of concept for the nascent digital ecosystem and proving that people would access some of the most important services in their lives online. At the same time, the investments made by the banking sector in digital identity authentication and cybersecurity, especially in developing a knowledgeable local workforce, would serve as critical resources as e-services took on a larger role in citizens' daily lives. In turn, investment by the Estonian government into telecommunications and digital infrastructure would create a flywheel effect as the more people who connected to the web meant more potential customers for the banks and subsequently more potential users of e-services.

With the first online banking services launched in the mid 1990s quickly showing their value, proponents of investment in e-government solutions were buoyed by the success of the sector. The time was ripe for the development of additional e-government services. The 1998 "Principles of Estonian Information Policy" had provided a high-level guide for the development of future services but there were still relatively few actual e-services to speak of. That began to change quickly in the late 1990s and early 2000s with the rapid introduction of an e-tax filing system, i-voting (sometimes referred to as e-voting), and an e-health system. While scores of other e-services have been developed over the last several decades, few have been as consequential as these three, notwithstanding digital identities and the X-Road which underpin and enable all e-services.

The e-tax solution launched in 2000 is one of the most important and widely used creations of the Estonian e-state given the importance of being able to raise revenues for government operations and social services. While by the new millennium the economy had been on the upswing thanks to the successful economic reforms put in place by post-independence governments, especially those of Prime Ministers Laar and Vähi, the country was still poor, especially compared to neighboring Finland where the GDP per capita was around three times greater.[9] The flat tax had been introduced in 1994 in a successful attempt to dramatically simplify the tax code and increase total collections,[10] and although collections were up, a state can always use more funds to deliver essential services.

Simplifying the tax administration process was a logical place to start. By digitalizing the service, the tax administration was able to drive down the time it took to pay taxes, creating a solution eventually used by 98% of the tax-paying population[11] and making the tax administration one of the most trusted public offices in Estonia.[12] The importance of decreasing administrative burdens imposed on taxpayers becomes clearer when one looks at the cost of compliance in other developed nations like the United States or Germany. In the U.S., the complexity of the tax code and the lack of a decent free government filing solution mean that Americans spent an estimated 1.7 billion hours and $31 billion doing their taxes according to a 2019 ProPublica investigation.[13] In Germany, businesses can expect to spend more than 200 hours annually on filing their taxes due to the complexity of the tax code and bureaucratic hurdles.[14] Unfortunately for larger markets where accounting and tax-related industries were more developed by the time e-government digital solutions became realistic, special interest groups have fought tooth and nail to maintain a status quo which benefits their business to the detriment of everyday taxpayers, a hurdle Estonia didn't have to overcome.

By developing an e-tax solution, the Estonian government hoped to not only increase the number of people who declared and paid their taxes but also to decrease administrative burdens. This would benefit both the taxpayers and the Estonian Tax and Customs Board which was tasked with managing the tax system, a goal that has clearly succeeded given the program's uptake. There was the added benefit that citizens didn't have to pay a private company to help file their taxes, compared to America where, according to reporting in 2024 by Vox, "Over ninety percent of returns are filed either by a paid tax preparer or by a taxpayer using commercial software, like TurboTax, which can cost as much as $169 per cycle, plus extra if you have state taxes."[15] However, in 2024, an estimated 140,803 taxpayers used the IRS' newly introduced free tax filing software, with the IRS estimating that this saved $5.6 million in tax preparation fees,[16] showing the potential benefits of such government-developed and run digital solutions.

Like the U.S., the Estonian e-tax concept started out small with a proof of concept (albeit more than two decades before the U.S.). The first stage of the project was launched in 2000 and enabled anyone to file their taxes online for free through a simple government portal. While this was useful and saved Estonians time and money by not having to hire accountants or use private sector tax software, the real breakthrough came later. In 2002, the e-tax system was expanded with automated tax declaration forms. In practice this meant that, "Using a secure e-ID, the taxpayer logs onto the system, reviews their data in pre-filled forms, makes the necessary changes, and finally approves the document with their digital signature."[17] Like many innovations in Estonia, they were not the initial inventor, but saw an idea or technology that they could effectively leverage and adopted it more competently than others—in this case after hearing about the Danish government using people's prior income return information to pre-fill their next tax return and then mailing it out. Instead of copying the Danish process wholesale, the Estonians decided to digitalize the effort, leapfrogging the paper-based concept.[18]

Despite the clear theoretical value of the system, adoption started off slowly with only around 3% of the population using it in the first year.[19] However, while taxpayers were initially hesitant, the government was able to quickly convince them with a major carrot: faster tax reviews and refunds for taxpayers who filed online, speeding up the process from its usual three to six months to under a month.[20] That incentive, combined with the fact that forms would come pre-filled with information already collected such as age, employment details, and dependents, made the service significantly more convenient than the standard paper filing process. This led to dramatically higher rates of use in the following years. According to 2023 data, 96% of individuals and 99% of companies now use the e-tax system.[21]

In the U.S., Representative Brad Sherman has called for the IRS to go "way beyond Direct File," saying, "Why in the heck should you have to fill out your return when the government has all the information…,"[22] in essence, advocating for an Estonian-like tax solution, showing its global applicability and value. Building on the success of the Estonian e-tax solution, the tax authority has continued to experiment with new technologies, especially those powered by artificial intelligence, to detect potential cases of fraud and to crack down on tax evasion,[23] far outpacing government agencies like the IRS which are just now truly dipping their toes into the digital era.

The e-tax system has become a calling card of Estonia's business- and citizen-friendly economic environment and is, by and large, uncontroversial. However, i-voting, another of Estonia's hallmarks, is more contentious. In 2005, the country became the first in the world to allow online voting in national elections,[24] in part thanks to the support of Prime Minister Laar.[25] In a digital country like Estonia, the expectation is that many government services will be online, but something as intrinsic to democracy as voting in elections is one of the few areas where citizens might be more resistant to moving the process online. This is doubly true considering the legacy of rigged Soviet elections which many Estonians experienced during the occupation. But in connected Estonia, the ability to vote online has now been available for two decades and has not experienced any proven cases of fraud.[26]

Arne Ansper, e-government expert and CTO of e-government services provider Cybernetica, compared the i-voting system to traditional postal voting where "An external envelope verifies the identity of the voter—a digital signature for internet voting—which is then stripped from the ballot, leaving an anonymous internal envelope guaranteeing the secrecy of the vote."[27] In its first year, just as with the e-tax system, there was apprehension in leveraging the new online system for something so important and fewer than 2% of voters used the system.[28] Things have changed significantly in the years since with more than 50% of ballots cast digitally in the 2023 elections. This marked the first time that the number of i-voters outweighed the number of people using traditional paper-based ballots,[29] showing how widely accepted and trusted i-voting has become across the populace.

The ease of use of the software (relative to voting in person or at an embassy for those living abroad) has led to some more humorous and adventurous uses of i-voting, including from an ice bath after a sauna session,[30] and often votes are cast from more than 100 countries around the world at election time.[31] Users of the system have skewed somewhat towards voters living in urban areas,[32] which are generally higher income than rural counties. This is in part also because of the agglomeration of tech companies in larger hubs like Tallinn and Tartu, where it has become a relatively common practice among such companies to encourage employees to take a short break to vote digitally during election season.[33]

With something as foundational and important to the functioning of a democratic society as voting in a free and fair manner, there was a great deal of consternation both in Estonia and internationally about moving any part of the election process online. While proponents argue that i-voting has made participating in the democratic process more equitable and accessible to a larger number of people, others have argued that the system is inherently flawed, exposing the electoral system to cyber vulnerabilities. It is the latter argument that has garnered the most attention in both domestic and international media, especially given the long history of animosity between Estonia and Russia. According to a 2020 interview with NPR, University of South Carolina computer science professor Duncan Buell claimed that "There is a firm consensus in the cybersecurity community that mobile voting on a smartphone is a really stupid idea,"[34] a feeling that extends for many in the cybersecurity community to any type of electronic voting. In an attempt to assuage fears over and promote transparency, the State Electoral Office of Estonia has made much of the source code available online via a public Github repository[35] and openly welcomes feedback from cybersecurity experts who spot potential vulnerabilities.

Frequently, the discussion around the pros and cons of i-voting moves to the idea of trust in the system. According to Tonu Tammer, the former Executive Director of the Estonian Information System Authority tasked with managing security in government computer networks, "Trust is the paramount factor in making sure that Internet-based voting actually takes place." However, according to Tammer, his greatest fear regarding attacks on the i-voting system is not from cyber threats but instead from fake news: "it's easier to erode trust by claiming electoral fraud than actually carrying out a successful attack."[36] This fear is not an unreasonable one, especially given the erosion of trust in the democratic process worldwide.

Within Estonia, some political parties have also stoked fears about the use of i-voting and potential vulnerabilities to the system, leading government officials to try to increase education on how the system functions and provide further transparency into the process.[37] However, there is also a clear incentive for certain political parties like the far-right EKRE, whose generally older voters by and large continue to use traditional paper-based methods for voting, to try to eliminate the use of i-voting, which mainly benefits the more liberal parties[38] who attract younger, more tech-savvy voters.[39]

The idea of i-voting has captured the imagination of many across the world disappointed with the lack of participation in elections and the lack of modernization of a decidedly archaic process. Bradley Tusk, who managed Mike Bloomberg's successful campaign for Mayor of New York City before becoming a tech investor and philanthropist, frequently opines about the importance of being able to vote online. He has even gone as far as to found a nonprofit organization focused on creating a mobile voting solution for the U.S,[40] fund pilot programs across the country,[41] and in 2024 authored the book *Vote With Your Phone: Why Mobile Voting Is Our Final Shot at Saving Democracy*. According to Tusk's nonprofit, the ability to make voting easier is existential for democracy: a broken primary system that leads to politicians being made "hostages" of the extremes of their parties has led to a system that doesn't work for the average American and, "The only solution is to dramatically increase turnout in all elections. That can only come about by making it dramatically easier to vote."[42]

The logic of the argument is clear, but in Estonia, increased ease of voting hasn't necessarily led to significantly increased total voter turnout according to both representatives of the Estonian government and a 2013 study on the topic.[43] Instead, it has been posited that i-voting mainly makes it easier for those who already would have voted to do so via the internet.[44] However, in narrow cases, i-voting has helped increase accessibility for those who may have found it hard to vote, like Estonians living abroad. In the 2023 election, the Estonian diaspora experienced a high turnout in part because of the ease of voting online, according to Atlantic Council.[45] Interestingly, Estonia has also struggled in other ways with citizen participation, despite attempts at developing easily accessible interfaces. This is best highlighted by the failure of TOM (*Täna Otsustan Mina*) which translates to "Today I Decide," an e-participation platform launched in 2001. According to the paper "Success in e-Voting— Success in eDemocracy? The Estonian Paradox,"

> The online platform, administered by the Government Office, allowed citizens to make proposals for new legislation and policies and discuss and vote upon them... Despite a relatively lively public interest in TOM, the project soon encountered challenges, such as a limited number of active users, low quality of ideas, limited impact of citizens' proposals and the prevalence of formalistic responses by officials over an open attitude to dialogue. By TOM's third birthday in 2004, e-democracy enthusiasts had declared it a failure.[46]

While the propensity of online tools to increase citizen participation in the democratic process can be debated, what is unquestionable is that i-voting has become part of the fabric of Estonia's society. According to a 2022 paper published in *Government Information Quarterly* which studied internet voting in Estonia from 2005 to 2019,

> The main finding of this article is that over a period of fifteen years, Internet voting has become normalized and even entrenched in Estonia. Electoral authorities no longer regard it as an experiment: i-voting is an essential part of the regular framework for conducting elections... The practice has been upheld by the courts and accepted by international organizations monitoring elections and democracy.[47]

While some parties and politicians that have not benefited much from i-voting have stoked fears about its use, the study found that around 70% of voters trust the use of online voting.[48] Given that in the 2023 elections more than half of all ballots cast were using i-voting,[49] voting online is likely to remain in Estonia as a safe, trusted, and convenient way to engage in the democratic process. And according to Professor Piret Ehin of the University of Tartu, domestic actors including the political elite have become invested in the continued use and success of online voting, leading to exceptionally high political and reputational costs for abandoning i-voting that would reverberate beyond election administration issues.[50]

With the health of Estonia's democracy in good order, the government was able to focus on another pressing problem for Estonian citizens—their physical health. Across the former Soviet Union, quality of life had stagnated during Communist rule and was still extremely poor relative to countries like Finland with which Estonia long aspired to be seen as a peer. According to official statistics from the Estonian government, in 2008 life expectancy in the country was 79.2 years for women and 68.6 years for men, increasing by just 3.5 and 4 years for women and men respectively over the decade post-independence.[51] This was poor compared to the European Union as a whole where women had an average life expectancy of more than 82 years and men expected to live for 76.1 years.[52] With the pressing problems of guaranteeing the country's sovereignty and putting the economy on the right track having been dealt with, the matter of poor health outcomes came to the fore.

Like many problems in Estonia, technology was tapped as part of the answer. Starting in 2005, the same year that i-voting was first used in an election in the country, the Ministry of Social Affairs developed a concept for what would become the e-health system.[53] At an incredibly rapid clip, the government worked with a variety of private sector partners as well as within the mostly government-run health system to develop the first iteration of the platform, with the main product being an interoperable electronic health record system which launched in 2008. By 2019, 95% of data generated by the healthcare system was digitized[54] and every person who receives medical treatment has a digital medical record.[55] Two years later, the government launched an e-prescription program which enabled doctors to easily send a prescription to a pharmacy—in under a year after launching the service 80% of prescriptions were issued using the new system and by 2021 that number had increased to 99%.[56]

Like many initiatives that would eventually fall under the "e-Estonia" banner, the idea of using information technology to make healthcare more efficient in the country originated outside of the government, as multiple local privately run health providers were already experimenting with electronic health records in the 1990s.[57] Seeing the importance of such systems thanks to the close collaboration between the public and private sectors, especially in the medical field where the government administered a huge amount of the healthcare system after the legacy of Soviet socialized medicine, the private sector's initiative quickly turned into one championed and eventually led by the government. The government had a major stake in making the system more efficient as providing healthcare took up roughly 4% of total GDP in 2008[58] via public spending. Any innovation that made the system more efficient and kept citizens healthier and more productive would therefore yield tremendous returns. At the same time, the use of digital tools that could automate pieces of the medical system or make doctors more efficient was seen as a way to alleviate another pressing problem facing the Estonian medical system: a lack of qualified doctors after many moved to Western Europe in search of higher wages.[59]

The electronic health records (EHR) system launched first and became the foundation of the e-health system. Despite the potential impact of the platform, it was granted an initial budget of only 1.6 million euros.[60] Comparatively, the U.S. Department of Veterans Affairs has spent nearly $10 billion since 2018 on a new EHR and costs could rise as high as $50 billion,[61] and attempts to implement a new healthcare IT system at Danish hospitals via an American software vendor at a cost of hundreds of millions of dollars were called "indescribable, total chaos."[62]

An EHR, in essence, creates an easily accessible digital record for each patient in the system with the intent of unifying every piece of relevant medical information in one place. Like the Once-Only principle, this means that when Estonians get a blood test, a dental exam, or a prescription, it is automatically attached to their personal electronic health record. When going to a local doctor's office, the physician simply accesses the patient's health record from the universal government database—accessible to all licensed medical providers through the e-health platform. All the relevant information is at their fingertips so they can focus on patient care instead of paperwork. The digitization of paperwork and digitalization of various burdensome processes results in tremendous time and cost savings for both patients and medical providers.

After Dr. Eric Topol, then chair of innovative medicine at Scripps Research, was asked during an event by a cancer survivor why they had to spend half an hour filling out redundant information every time they go to a doctor's office, Topol responded emphatically about the urgency of the issue and recommended that the U.S. look to Estonia as an example.[63] An *Atlantic* article discussing the event reported,

> The American medical system is atrocious at keeping track of the stuff it does. According to Topol, 10 percent of all scans in the United States are repeated unnecessarily simply because patients can't get hold of their past records and scans. It amounts to billions of wasted dollars.[64]

The EHR solution developed as the foundation of Estonia's e-health system meant tremendous efficiencies both for the medical system and for patients. No filling out redundant forms, no repetitive tests due to lost paperwork, and no delays in getting needed care when doctors had every piece of information in one place. Additional services and integrations were also made with the EHR, from e-prescriptions to e-ambulance services. With e-prescriptions, the idea is conceptually simple—a doctor could now send a prescription digitally to a pharmacy to be picked up—but comes with enormous benefits. For example, when someone comes to pick up a prescription, pharmacists can provide more information to the person (like whether they are eligible to participate in screening programs for certain types of cancer based on their health history or whether there may be an issue with various drug interactions). This works no matter the pharmacy the person goes to or whether they have been there before, helping make the healthcare system more proactive in supporting citizen health rather than simply reacting to health crises as they arise.

The e-health platform has also been a literal lifesaver in emergency situations. With the e-ambulance system integration, when someone calls 112 for a medical emergency, they can provide the digital identity number of whoever needs help to the dispatcher. With this information the digital identity is transmitted to the ambulance along with the requisite health information so that when they arrive, paramedics know if the person is on certain medications, has allergies, or has a history of certain diseases so that they can deliver the right care, faster. If the situation is serious enough to require a hospital visit, the paramedics in the ambulance (or the dispatcher on the phone) can add in new details to the patient record so that as the ambulance is in transit, the hospital can prepare for whatever is coming.[65] In critical emergency medical situations like strokes or heart attacks where seconds count, e-services can save lives.

While the idea of such e-health products may seem emblematic of an advanced healthcare system that prioritizes patients and efficiency to some, for others the concept is fraught with privacy risks. When fielding questions about the e-health system from foreign journalists, concerns over patient privacy are likely to come up, not without reason given that medical data is likely the most sensitive data a person has. Fears over privacy risks are increasingly common given numerous data leaks, hacks, and more which expose patient data, and while Estonia's government has a strong track record in keeping data safe, there are valid concerns over the collection of significant amounts of health data. The potential for the misuse of data must also be seriously considered, like whether a more corrupt government could sell data off to pharma companies over the concerns of citizens or from basic privacy violations by government employees peering into private medical information. However, these misuse scenarios are less a technological issue than a governance issue—the Estonia digital government infrastructure has proven itself safe and secure, but if rule of law were ever to crumble, then health (and other) records, now easily accessible, could be weaponized or monetized by unscrupulous actors.

The leadership of the Health Ministry naturally takes a significantly rosier view on the use of digital health solutions and their benefits. According to a 2019 interview with Minister of Health Riina Sikkut,[66]

> For the patient, digital health solutions, particularly in the era of GDPR [General Data Protection Regulation], are a way to take control of their own data. This has long been a cornerstone of the Estonian e-health system: patients can access their data, see who else has accessed it, and also close off their data from the system.[67]

BECOMING E-ESTONIA

And according to Sten Tamkivi,

> This liquid movement of data between systems relies on a fundamental principle to protect the privacy of the citizens: without any question, it is always the citizen who owns their data. People have the right to control access to their data. For example, in the case of fully digital health records and prescriptions, people can granularly assign access rights to the general practitioners and specialized doctors of their choosing.[68]

Because the e-health platform uses the X-Road as its backbone for data exchange and digital identity for authentication,[69] user privacy is built in from the start and everyone can see who has viewed any of their documents or can proactively lock certain records so that they can't be viewed by anyone, and can take legal action if they do (although this is dependent on laws being enforced and the government not surreptitiously creating backdoors). It is also because of the X-Road that the e-health platform can seamlessly integrate with other government services, like a doctor being able to submit required vision and health checkup information to the national road administration directly from a patient's electronic health record for a driver's license renewal.[70] Often, these integrations are some of the most popular services for end users.[71]

As with other e-services, the government has worked to create clear legal rules for their development and e-health is no exception. In 2001, The Health Services Organization Act was proactively introduced with guidance for patient protections and clarity for healthcare providers in anticipation of the development of e-health services. According to the study, "Ten Years of the e-Health System in Estonia," from the Tallinn University of Technology's Department of Health Technologies and Information Systems Group,

> The Act also states that access to patient data is available only to licensed medical professionals, legal representatives or patients trustees... The Act also states (and this is realized in the patient portal), that the patient has the right to hide their data so that healthcare professionals are no longer able to view them.[72]

The Estonian legal system has also worked to try to save money for both consumers and the government by enacting simple requirements such as mandating that when someone goes to pick up a prescription, they are offered the lowest-price version first if there is a generic alternative available.[73]

Estonia is now widely recognized as a leader in digital health thanks to its ongoing innovation in making services interoperable, its proactive healthcare interventions, and its investments in, for example, precision medicine via free gene tests.[74] As such, the country has been tapped to help groups as diverse as the World Health Organization[75] and the Moldovan government[76] with the development of e-health technologies. According to a 2019 Bertelsmann survey undertaken before the global COVID-19 pandemic, Estonia was ranked number one on their digital health index, above much wealthier and larger nations like Canada and Germany,[77] and the country's reputation as an e-government leader has led to fawning articles in the international press.[78]

Despite Estonia's leadership in digital health, the country still faces serious challenges when it comes to holistic healthcare and life expectancy, with some believing that the promise of the digital health system has been oversold. From 2000 to 2017, life expectancy for both men and women rose dramatically to an average of 78 years[79] at a pace faster than in any other EU country during the period,[80] but that number still trailed the EU average.[81] The country also faces major disparities between the minority Russian-speaking population and ethnic Estonians, a long-running issue in a range of socio-economic areas.[82] At the same time, there is a large gender gap with women expected to average an additional nine years of life compared to men.[83] Despite investments in e-health services, the overall health system is funded at less than half the EU average, leading a 2021 European Commission report to call the Estonian healthcare system "chronically underfinanced."[84] The COVID-19 pandemic also laid bare gaps in the overall healthcare system that could not be filled in by e-services.

Dr. Keegan McBride, a University of Oxford Lecturer on AI and government policy among other issues who has been intimately involved in Estonia's e-government operations for years, has cautioned against seeing e-health services as a panacea. When asked how Estonia's digital prowess had helped it deal with the COVID-19 pandemic in an interview with *Estonian World*, the country's main English-language news publication, he answered:

> Yes, some parts of the e-state worked phenomenally. A clear example of this is the amazing work that TEHIK (the Estonian Health and Welfare Information Systems Centre) has done to make data available in an open format—Estonia is probably a leader when it comes to making COVID-19 data available. But, in many other regards, the e-state did not help. The primary reason for this is that digital tools cannot be a replacement for experience, capacity, and capability.[85]

Between the disparity in the healthcare system regarding the average life expectancy for men and women and Russian- versus Estonian-speaking peoples as well as serious complications regarding Estonia's response to the pandemic, it is clear that while impressive, e-health can't solve everything. According to McBride, the government should learn that "the e-State is not a substitute for good governance, one must have both."[86] Even the e-state had issues during the pandemic, with two senior government officials discussing the failure of the vaccination booking system in a state-run blog and acknowledging that the system (briefly) ended up inadvertently blocking many citizens from booking vaccinations because of an unfriendly user experience.[87]

Despite its flaws, Estonia's e-state is the envy of much of the world and the subject of a great deal of global reporting and fascination. This includes President Obama, who said, "I should have called the Estonians when we were setting up our healthcare website."[88] Citizens can file their taxes in a few moments, vote from their laptops during a coffee break, and know in a healthcare emergency that doctors will have the information they need to make the right decisions for their patients.

However, the initially slow rates of use of many of these technologies show that the Estonian people are not de facto techno-optimists but instead saw the value of each service before adoption. According to renowned venture capital investor and author

Ben Horowitz, "Now independent, but well aware of their history, the Estonian people were humble, pragmatic, proud of their freedom, but dubious of overly optimistic forecasts. In some ways, they had the ideal culture for technology adoption: hopeful, yet appropriately skeptical."[89] And e-services, while important, clearly cannot be considered a universal remedy for every problem in society—instead, they must be considered as part of a larger solution. According to Wolfgang Drechsler, Professor of Governance at the Tallinn University of Technology and Honorary Professor at UCL, digitalization can only amplify existing state capacity: "What you can learn from Estonia is that digitalisation is like MSG, it enhances the flavour but it depends on what is already there."[90] But e-services have no impact at all if people can't access them or understand how to use them. It was by investing in citizen education to teach individuals how to effectively leverage the web and the power of computing that Estonia was able to become a digital society, something which was inspired thanks to the Mosaic browser developed by Ben Horowitz's colleague, Marc Andreessen.

8

PLAYING LEAPTIGER

The opportunity for Estonia to compete in a new arena where others had fewer relative advantages appeared unexpectedly in 1993. That year, Eric Bina and Marc Andreessen, students at the University of Illinois at Urbana-Champagne, released the Mosaic browser.[1] Andreessen is now a well-known investor and the "Andreessen" in the storied venture capital firm, Andreessen Horowitz.[2] Mosaic is widely known as the browser integral to making the internet accessible to the average person through an interface that was easy to use, install, and understand, managing multiple internet protocols rather than putting the onus on the user. A decade later, Skype would use a similar formula to triumph over its competitors by making the product accessible and easy to use even for someone with no technical knowledge.[3]

Just after the browser was released, Toomas Hendrik Ilves began serving as Ambassador to the United States (some years before becoming President). He had been exposed to software development at a young age and worked as a programmer in college, becoming significantly more technically proficient than the average politician. After downloading and using Mosaic and seeing how early the nascent internet economy was in its development, he understood that for Estonia, "this is someplace that we are on a level playing field. It doesn't matter the U.S., Germany, Sweden,

Finland, no one has a real leg up in this. They are at the same place as we are."[4] To others, the internet was more than an area where Estonia could thrive economically: it was a place where they could freely access the world after decades behind the Iron Curtain. According to Linnar Viik, "for young Estonians, the internet is a manifestation of something more than a service—it's a symbol of democracy and freedom."[5]

Finally, Estonia had an opportunity to compete in an area where they were equal, because everyone was starting from essentially the same place. But, for Estonia to capitalize on the new innovations happening in the IT sector, they needed to dramatically shift gears. With the digitalization of the economy already on his mind, Ilves was searching for a way for Estonia to leapfrog from the current economic situation to one more comparable of that of neighboring Finland. He recognized that all successful countries had one commonality: a highly skilled and technical workforce.[6] According to Henry Kissinger, Lee Kuan Yew had also identified the importance of creating a skilled workforce for achieving his dream of a successful, sovereign Singapore despite its lack of natural resources, saying: "his vision was of a state that would not simply survive but prevail by excelling. Superior intelligence, discipline, and ingenuity would substitute for resources."[7] Singapore's eventual success highlights the potency of such a policy prescription and the similarities between the nations.

While society-wide educational reform was a multi-year undertaking, with a decade or more liable to pass before seeing any tangible returns on the investment, Ilves and many key Estonian leaders were open to radical reform and were ready to invest in the nation's long-term future. With the web newly accessible thanks to the Mosaic browser, there was now the possibility of giving isolated Estonia a virtual passport to the world and a level playing field for it to compete on. The internet had the added benefit of helping the country overcome its geographic limitations, namely being relatively isolated from large Western markets (except by sea), leveling the playing field further for the country as they competed for resources, sales, and investment with larger and more centrally located nations. But for the country to truly benefit from the economic reshaping beginning in the nascent digital era, they needed the tools and education to seize on the internet's potential. For that, large-scale educational reform and infrastructure development would be required.

A former journalist, Ilves retained his investigative and literary skills and remains an incisive communicator. To create a program with the ambition of doing nothing less than completely reshaping the Estonian society and economy, he knew that he had to plant a stake in the ground and create public support. He did so by first pitching the idea of what would become the Tiger Leap campaign—a project that would provide computers and connect every school to the web while teaching digital skills to both students and teachers—to the Minister of Education as well as discussing it with the President, Prime Minister, and the banking sector, all of whom were at least rhetorically supportive. In early 1995, an article about Ilves' proposed program including a preliminary estimation of its price appeared (of $10–15 million for 17,000 computers) in *Rahva Hääl*,[8] one of the major newspapers. It then quickly began making the rounds within the academic community via an online electronic mailing list, as academics were one of the few groups in Estonia who had access to computers at the time.

Many were supportive of the proposal, but it was not well received by all Estonian society. According to Ilves, "Tiger's Leap was not always popular. The teachers' union had a weekly newspaper, and for about a year, no issue would seem to appear without some op-ed attacking me."[9] However, with the benefit of hindsight, it is clear that Ilves was correct in championing the Tiger Leap program. Tiger Leap, along with future digital literacy programs taken up like the Look@World campaign which was supported by the financial services, telecom, and IT industries, led to a rapid increase in the number of internet users, with the goal of making Estonia the most connected nation based on internet users per capita by 2003. At the same time, the rise in internet users would create more customers for products like online banking and the programs led to a dramatic increase in cyber literacy in the country while creating a talent pool with the skills needed to develop both a digital government and a thriving tech ecosystem.

Thanks in part to the foundation laid by the Tiger Leap, as of this writing, the country boasts the most startups, unicorns, and investments per capita in Europe[10] and 93.2% of households are connected to the internet.[11] But in fairness to Ilves' detractors, like those in the teachers' union who feared spending money the country couldn't spare on a wild idea, they were being completely rational. The Estonian economy was on the upswing in the mid-1990s but it was certainly nowhere near the point where there were discretionary resources to spend on moonshot ideas. But for Ilves, it was a bet worth taking and, fortunately, he found a champion in Dr. Jaak Aaviksoo.[12]

A physicist and educator, Aaviksoo was no stranger to the ongoing advances shaking up the technology ecosystem. In Estonia, as with much of the world at the beginning of the digital era, university professors and students were often the first to be exposed to computers. Aaviksoo inherently understood the importance not just of connecting schools, but making sure that the next generation of Estonians had the technical skills required to succeed in the emerging digital era, saying "the establishment of at least one internet station in each of the country's 254 municipalities was as important as joining the EU and NATO."[13] While Aaviksoo had a storied career in academia, he also had the political skills necessary to navigate a complex political landscape and helped to shepherd the fledgling idea into a concrete program that had more than rhetoric and bold ideas behind it.

Together, Ilves and Aaviksoo continued the development of the concept laid out in Ilves' 1995 proposal. With a temperament for taking bold action, Aaviksoo swung into full gear in support of the idea, working to find partners who could help supply and finance the equipment and material that the program would need. Soon after, they brought together a group of similarly passionate advocates for reform. Along with Ilves, there were engaged citizens like Ants Sild, Tanel Tammet, and even Sten Tamkivi, then a student in high school who served as a youth advocate on the committee. For Aaviksoo and Ilves, the group's main purpose was to solidify the specific goals of the program and to make sure that it moved from idea to reality.[14] But even with a strong advocacy group and champions like Ilves and Aaviksoo, the Tiger Leap was such an ambitious idea that it needed political backing at the highest levels for a chance at success.

Thankfully, the group found a true patron in Lennart Meri, the long-serving statesman and much lauded writer and director. Meri had played a storied role in Estonia's independence movement, acting as a voice of the people, daring to say that which most could or would not, often at great personal risk. His work as the country's foreign minister immediately after independence had solidified his near legendary status not just in Estonia but in diplomatic and press circles across the world, where many called him "Mr. Estonia," due to his personification of a country few (at least in the United States) knew much about.[15] In 1995, Meri was serving as President of Estonia, a role he would continue to hold until 2001. Like Aaviksoo, Ilves, and scores of others, he recognized that for the people of Estonia to succeed, radical reform and bold political leadership were required.

On 21 February 1996, just a few years after the release of the Mosaic browser, President Meri publicly announced the Tiger Leap program to the Estonian people.[16] The program had one overarching goal: to connect all schools in Estonia to the internet before the new millennium. While one component of the program was providing schools with computers, the key focus was always on connectivity. For Ilves and others, it was connectivity both to the world and between different parts of Estonian society that would be what truly disrupted the Soviet legacy. Their instincts were proven true not just thanks to the success of Tiger Leap kids, many of whom became founders and early employees of top local startups like Skype, Bolt, and Wise, but through the connection of teachers across the country. Once online, teachers were empowered to share best practices and build their own curriculums as well as digital tools. Today, the country is a global education leader as the top European country in 2022's PISA rankings (along with Ireland).[17]

For a country with a significant rural population and several sparsely inhabited islands, connecting every school to the web was no easy feat. For Estonia to have a chance at meeting this ambitious goal, they needed to not only galvanize the support of the public and senior political leadership but ensure that the program was structured with incentives in place that would give it the best chance of success. To do so, rather than put the program within the Ministry of Education, as might seem logical for a program intrinsically linked to children's education, they created a foundation that would be responsible for executing the vision of the Tiger Leap campaign.[18] According to Aaviksoo, when the foundation was set up it was made clear that it would not be funded or run solely by the state but also would depend on contributions from companies and individuals. It was to be a bottom-up initiative rather than one dictated top-down, which he credited as at least part of the reason for the Tiger Leap's success.[19] Ants Sild, one of the founders of the Tiger Leap Foundation who has continued to work in ICT capabilities development projects in Estonia and around the world, is inclined to agree. However, he also highlights the importance of involving all related stakeholders like IT companies and specialists, teachers, local communities, and citizens who wanted to be supportive of the program and contribute their time and knowledge to help develop the program's strategy and execute its ambitious agenda.[20]

This blending of public and private would be one of the more consequential decisions in the initiative. By helping to bypass the regulatory and administrative morass of government while still being able to count on senior government officials, the foundation had the space it needed to move quickly and experiment with how to best execute their ambitious mission. The foundation's structure had another perk—it could raise donor funding and form unique partnerships to support the initiative, including from local IT companies and their global partners as well as international institutions and NGOs like the Open Estonia Foundation. However, the vast majority of funding came out of the Estonian government's budget. Building synergistic public-private partnerships would continue to be a superpower of the country, especially when it came to connectivity related issues. The Look@World campaign, which eventually helped to teach around 10% of Estonia's adult population digital skills according to the e-Governance Academy,[21] was spearheaded by Hansapank (now Swedbank) who provided funding for the initiative[22] along with significant support from other local banks, telcos, and IT companies.

When the Tiger Leap foundation formally launched along with Estonia's first public internet access point in February 1997, it marked the beginning of Estonia's sprint towards becoming a "unicorn country" where connectivity and opportunities in the digital world abounded. For the founders of the program and the scores of others who helped to make it a success, there was no time to waste in helping prepare the country for a connected world. Enel Mägi, who joined the Tiger Leap team as the lead administrator after working at the Open Estonia Foundation, helped move things into gear. In Mägi's words,

> The Tiger Leap Program is a step toward ensuring our success in competing with larger nations in the 21st century, when the world is evolving into a society in which information is the main commodity. Estonia is willing to invest in the future of its people.[23]

To ensure the program would be successful and embraced by all Estonians, Mägi and the Tiger Leap team executed a complex, multi-pronged strategy.

One of the first areas of engagement was building public support through educating people as to just what the computers and the internet were capable of. While some academics, university students, and businesspeople had been exposed to the web, the average Estonian, including most schoolteachers, likely had not. Within a few months, the Tiger Leap team brought the public's attention to the power of computers by launching a free event at the Tallinn Town Hall square.[24] There, they set up computers that anyone could come and explore, giving the average person access to something they likely could not afford and showing them the potential of the technology. The tactic of publicly demonstrating the power of computers and the web would be a powerful one in the Tiger Leap team's repertoire, helping them build support with both the public and key stakeholders who were unfamiliar with computers and their potential benefits. Within a short period of time, the team helped to organize events in numerous Estonian counties where the public could come and play with the computers and participate in various competitions and lectures. The use of broad societal education as a catalyst for change in the country is a tool that would be used repeatedly by the public and private sector, including with the Look@World campaign which initially focused on digital literacy and would eventually expand to cyber hygiene and other relevant skills.[25]

The second prong of the Tiger Leap team's strategy was to create incentives that would not only convince teachers to go along with their vision, but would also create alignment between the government, the foundation, and schools. To do so, only a few months after the official launch of the foundation, a program was launched that would help teachers purchase discounted home computers by offsetting part of the cost.[26] The idea behind the initiative was that if teachers who were already busy with their existing teaching responsibilities were less likely to actively seek out learning opportunities on computers and the nascent internet, then the Tiger Leap team would have to make it even easier for teachers to access the web and see the power it held. This subsidization, paired with the public demonstrations of computers, helped to dramatically increase adoption by the teachers who would soon be tasked with educating their students according to a new IT-focused curriculum meant to build the skills necessary for the development of a tech-literate workforce. At the same time, Ilves had always believed that the program should be run in a way that mandated that every party should put at least some skin in the game so that they truly understood the value of the equipment they were receiving. According to Ilves, "if you just give people computers nobody is ever going to use them,"[27] so programs were structured in a way that promoted cost-sharing between parties, with the federal government and companies helping make IT equipment accessible by offsetting part of the cost, but with local governments and teachers expected to pitch in from their own budgets as well.

The third prong of the Tiger Leap campaign was providing training and support for teachers and school staff while working with global partners to create a curriculum fit for a rapidly modernizing society. While Estonia boasts a difficult language and many locals spoke some English thanks in part to their exposure from subtitled Finnish television broadcasts of American shows (something not shared with the Baltic nations of Latvia and Lithuania which were too far away to receive TV signals from Helsinki), the team knew that for the sake of accessibility they would have to create localized training and curriculum materials for teachers and students. But rather than building from scratch, the Tiger Leap team discovered that Intel had already developed a training course like the one the country needed.[28]

The team was able to cooperate with Intel and localize the company's "Teach to the Future" training program[29] for the Estonian context, allowing the team to skip a major step and expense in the process. In the first year alone, nearly 4,000 teachers participated in the free course the team put on[30]—a significant number given there were fewer than 20,000 teachers at the time. Even more impressive, when the impact of the program was assessed, it was shown that by early 2001 more than half of Estonian teachers had received basic training and 2,600 of them had participated in an advanced course.[31]

The Tiger Leap initiative had started with one goal above all others: to get all Estonian schools connected to the internet before the start of the new millennium. By the end of 1999, they had accomplished that goal. It wasn't always pretty—many schools didn't have top-notch connectivity or the most sophisticated computers, but they all had a baseline level of access.[32] Like Ilves had imagined, every Estonian student now had at least the same opportunity to compete in the emerging digital economy as their Western counterparts (sometimes better opportunities given the still gaping digital divide in the United States, with tens of millions of Americans lacking access to high-speed internet as of 2022),[33] something that was unthinkable just a decade earlier.

For Ilves, Meri, Aaviksoo, Viik and the scores of others whose hard work brought the vision of connecting the country to the nascent web, the program was a success. By involving the private sector, aligning incentives with teachers (and creating new incentives), working to build up public support, and creating localized trainings and curricula, the Tiger Leap team managed to achieve an unthinkable goal in only a few years. Now, most teachers across the country had been exposed to the power of computers, children had access to the web and could see the world and all its possibilities through its browser, and the Estonian nation was on its way to building a well-educated workforce with the right skills for the digital era. Many others in the Estonian government and civil society seemed to agree given the number of similarly named spin-out programs that would emerge in the country, including Science Tiger and Tiger Robotics.[34]

For the youth in Estonia, many of whom remembered living under the shadow of the Iron Curtain, their new education and access to digital infrastructure gave them something more than opportunity: it gave them hope and let them dream big. According to tech entrepreneur Ott Kaukver, the "Internet broke down the boundaries, the borders of countries. Now you could do a global thing while sitting in Estonia."[35] For a people long physically distant from most major economic markets, who had been violently separated from the rest of the world by the Soviet's Iron Curtain, being connected virtually opened a world of possibilities for commerce and innovation, enabling the development of a country ripe for unicorns and innovative disruption.

9

THE ESTONIAN TECH MAFIA

Hanging on the wall of Lift99, Tallinn's best-known coworking space and hub for all things tech in the capital, are gold plaques with the names of startups that have joined the Estonian tech mafia's "wall of fame."[1] A tech mafia, as it is commonly called in the startup world, refers to the founders, early employees, and core investors of a successful tech company like PayPal and Palantir and doesn't carry any negative (or violent) connotations. Like the mafia, Estonia's tech community is extremely close-knit and obsessed with supporting their compatriots (and the country itself), in part due to what one founder called an irrational patriotism, especially among those who still remember the Soviet occupation.[2] Once you're in, the tech mafia often feels more like an extended family than a business network. Estonians abroad band together as members of an elite club of just more than a million who speak the same language and have a shared culture that prioritizes lifelong relationships. According to entrepreneur-turned-investor Sten Tamkivi,

> I think that this has its roots in coming from a tiny country. There is no local market in Estonia, or at most a very small one: you cannot build a billion-dollar company focusing on a market of 1.2 million people. What that means is that everything you start has

to be global. So whatever you are building, you never presume one language, one currency, one nationality, one culture. This creates a tight and collaborative community, as we are all looking outside of Estonia. When a young Estonian founder is looking to raise their first seed money there are entrepreneurs ready to be angels, there is direct access to networks. The eco-system is there, and it is very supportive. As founders, the only thing that we compete on is talent.[3]

It sounds contradictory, but the insular and close-knit nature of Estonia's population has allowed for a uniquely open attitude towards commerce. Because of its insularity, selling products or raising local capital in Estonia is difficult for foreign companies and founders, creating a unique advantage for local entrepreneurs who can start small locally with the intention to scale globally. Indeed, building products with the idea of scaling to foreign markets rapidly is a necessity as the domestic market is too small to support any high-growth startup for long. However, thanks to the Estonian tech mafia, entrepreneurs know they will have the support they need to scale and can think big from day one.

In Estonia, the mafia that started with the founders and early employees of Skype has snowballed into a much larger group since the company was purchased by eBay and eventually by Microsoft. According to the investment firm Atomico, founders and employees of Skype are responsible for creating more than 200 companies[4] and it is near impossible to find an Estonian startup that does not have a founder or investor that was associated with Skype in some way. For instance, before teaming up in 2021 to create a new investment fund,[5] Taavet Hinrikus and Sten Tamkivi were both early Skype employees and friends since high school and had made dozens of angel investments independently.[6] Taavet went on from Skype to build TransferWise, now a multi-billion-dollar publicly traded company, and Sten became an Entrepreneur in Residence at the storied venture capital firm Andreessen Horowitz before founding a startup which was subsequently acquired in 2017.[7]

Newer Estonian startups like Bolt, an Uber competitor with operations around the globe, were helped along greatly by the Estonian tech mafia. Multiple early investors in the company hailed from Estonia[8] and the company garnered an eventual valuation of more than $8 billion,[9] making its CEO, Markus Villig, the youngest founder of a billion-dollar company in Europe at the time.[10] According to an op-ed in *Politico* in late 2022, the ICT sector has blossomed into a major driver of jobs and tax revenues, with 8% of total GDP coming from the industry and early-stage tech start-ups employing 1% of the country's workforce, plus an anticipated 25% annual growth rate for tech exports.[11] And venture firm Atomico's tenth anniversary edition of their State of European Tech report published in late 2024, claimed that "Estonia takes the world's top spot by the relative scale of investment in tech, with VC funding accounting for 1.17% of GDP over the past decade."[12]

While discussing their new investment firm in 2022 with Amy Lewin, the editor of a *Financial Times*-backed publication, Sten and Taavet explained how they see their legacies and what they owe the next generation while providing unique insight into the culture of the Estonian tech mafia.

> "There's a mentality of giving back," says Tamkivi—and it's not just visible in Estonia. The Finnish gaming ecosystem has a similar dynamic, with Nokia giving birth to Supercell and Rovio. "It's a virtuous circle." Investing in tech isn't all that Hinrikus and Tamkivi do. They're also hugely influential in the Estonian tech scene—and wider society. Tamkivi was president of the Estonian Founders Society for three years, sits on boards of museums and financial groups and is a member of the council for Estonia's famous e-residency programme.[13]

Tech mafias are nothing new. They've emerged from PayPal, Google, and scores of other top tech companies and universities from Silicon Valley to Helsinki as former founders go on to become the investors and founders of new companies and support one another in their new ventures. However, in Estonia, the tech mafia benefits from an additional supporter—the government. Rather than look at the tech sector as a distinct entity or as just another tax revenue opportunity, officials have long seen the private sector as a key part of how Estonia would develop into a globally known brand and power. For the private sector, the support of the government is critical for their businesses. This was doubly true in the days before Skype's success when the public perception of Estonia among those who knew much about the country was often negative. According to Arne Ansper, CTO of Cybernetica, an e-services company which sells its products and services across the world, "the reputation of Estonian companies was 'a small, suspicious, Eastern European country with strong links to the Soviet Union,' and it didn't earn trust in many fields."[14]

Because Estonian tech companies weren't often given the benefit of the doubt of being able to deliver on lofty promises, especially relative to contemporaries in Silicon Valley, the government often became a "proof of concept" customer to help them along. This gave companies a leg up, or at least a leg to stand on, as they competed abroad by being able to show they had a product that was already working. For the government, having a tech sector with a global reach was just as important. Skype helped to put Estonia on the map in the minds of many, creating a tech-forward, progressive image for the country. Startups ranging from Skeleton Technologies (which invented a new type of supercapacitor critical to the green transition and has raised more than $300 million as of 2024)[15] to Starship Robotics (which creates autonomous delivery robots available across the U.S. and has raised more than $200 million from global investors)[16] continue to serve as examples of Estonia's engineering and startup prowess, all while creating jobs and tax revenue.

Given that the public and private sectors understand one another and are in frequent contact, they also can work to support one another. Sten Tamkivi has said "I can actually send an email to the President, I can actually send a text message to the Prime Minister and get answers."[17] The CTO of Pipedrive, a leading Estonia tech firm last valued at more than $1.5 billion when a majority of its shares were purchased by a private equity investor,[18] has also highlighted the close connections between the tech community and government, sharing that at least once a year the Prime Minister holds a roundtable with the community to talk about shared issues that need to be addressed.[19]

Minister of Economic Affairs and Information Technology, Tiit Riisalo,[20] also emphasizes that the government actively tries to understand the startup sector, saying, "we talk with people who are running the companies, doing the startups, so we know the needs and try to reflect them."[21] The longtime strategy of using private companies to build e-government systems along with an unrelenting focus on supporting the international promotion of those same companies shows how seriously the government takes their role in supporting the private sector. But the idea of the critical symbiotic relationship between both public and private sector predates most contemporary Estonian politicians and came from Mr. Estonia himself, President Lennart Meri. A man well known for his aphorisms and keen observations, Meri would ask, "What is our Nokia?"[22]

For a country somewhat obsessed with catching back up with its neighbor, Finland, it was a profound question and one used to push colleagues and contemporaries in the public and private sector to create a strategy that would put Estonia on the proverbial map after independence. Meri had seen that for Finland, Nokia served not just as a large source of employment and tax revenue (accounting for about 4% of Finnish GDP in 2000 during its heyday),[23] but also as a calling card for the country. Nokia was a global symbol of local engineering prowess and Nordic design, and a living demonstration of Finland's relevance on a global stage usually dominated by significantly larger countries. Meri knew that for Estonia to stand out despite its miniscule population they needed to have an answer to his question, and soon. For much of Meri's time in office, the rhetorical question did not result in concrete action by the government[24] given the number of other pressing challenges that had to be dealt with, but eventually the answer to his question would appear thanks to Estonia's first unicorn, Skype.

Thanks to several already implemented and ongoing reforms, in addition to the passionate and patriotic support of the new Skype mafia, the country was in an advantageous position relative to its peers and even much larger and richer nations. The flat tax system for individuals and businesses meant that companies weren't crushed by onerous tax obligations while they were getting off the ground. The government listened to entrepreneurs when they came across issues unique to building up the startup sector, working to provide clarity and introduce new legislation when necessary to enable the industry, like changing how share options were taxed to make the system more globally competitive.[25] This is just one of many regulatory changes that have helped land Estonia a top spot in the ranking of European countries by Not Optional, an advocacy group led by Index Ventures in collaboration with hundreds of CEOs "with the aim of improving the competitiveness of European startups."[26]

The e-services that had already been launched, ranging from online tax filing to the e-business registry, meant that it was painless to start and manage a company and required little managerial overhead. According to Capability Development Director Raul Rick at Milrem Robotics, one of the leading Estonian defense tech companies, Estonia's digital governance has been key to their success, as he says, "The country's leadership in e-services, like digital ID and electronic signatures, has streamlined our operations, making daily administrative processes more efficient. This digital infrastructure has been a significant enabler as we innovate and expand."[27] And because e-services helped to eliminate corruption to a significant degree, there were no impediments to building a successful business from the mafia (the actual mafia, not a tech mafia) like in neighboring Russia where growing companies could be shaken down with often realized threats of violence at any time. Investments in connectivity across the country meant that the costs to get online had been driven down for consumers and businesses and there was not much of a "digital divide" to speak of, enabling people to access the internet and build a global company from most anywhere in the country, which had the side benefit of facilitating remote work for many.

However, one of the most important factors behind Estonia's competitive advantage as an emerging tech hub was education. The Tiger Leap campaign connected schools and helped to prepare youths for life in a digital world, giving them the skills that they would need to build successful companies in the online economy. More than that, the country prized high-quality education and the learning of useful skills from English to robotics and programming, so when new tech companies began to grow in the country, they

THE ESTONIAN TECH MAFIA

found a pool of knowledgeable and capable workers at the ready. According to TransferWise founder Taavet Hinrikus,

> Estonia has a highly developed digital economy and there are a large number of smart people and given that both me and my cofounder are from Estonia, we knew lots of people here and it was pretty obvious for us to start tapping into the people we know here to help us grow TransferWise.[28]

But becoming a startup ecosystem is no easy task, even with a few meaningful advantages, especially when major existing tech and financial hubs like Silicon Valley and London were competing for talent and had significantly more established ecosystems and resources to bring to bear.

Like they had many times before, policymakers leaned into Estonia's unique environment and played to its strengths: openness, an ability to move faster than other competing jurisdictions and implement common sense reforms, a close-knit culture, and strong links between the public and private sectors. There are two programs that are now emblematic of Estonia's drive to become a leading tech hub, Startup Estonia and Accelerate Estonia. Startup Estonia was formed in the mid 2010s with a mission to support the development of the domestic startup ecosystem.[29] The organization serves as a vital link between the tech and government communities, working to advocate for the needs of the tech ecosystem and ensure understanding between both. Accelerate Estonia was created several years later with the idea of facilitating the development of moonshot ideas in a much more hands-on way, operating as something more akin to a startup accelerator program than a government agency. Also critically important are an array of other government-funded projects to support the startup ecosystem, including SmartCap, which served as a fund of funds for privately run Estonian venture firms, and Tehnopol, a science and business park, which has become a hub for early-stage technology incubation and a partner for startup acceleration programs like NATO's dual-use startup accelerator, DIANA.

Startup Estonia has spearheaded and championed projects large and small, from the development of an ever-popular set of standardized common legal documents for startups[30] to help them save on legal fees, to supporting the development of a new visa class in 2017 to bring foreign startup founders to Estonia.[31] Their efforts have led to serious changes for the startup ecosystem in the country. According to the agency's 2020 White Paper, the Startup Visa program alone has led to more than 500 founders of potential high-growth startups moving to Estonia from across the world,[32] although there have been some difficulties integrating foreign founders into the tech community and insular Estonian society along the way. Their efforts to make the country a friendly place to tech entrepreneurs and technologists along with those of the government's talent attraction team at Work in Estonia also meant that as of 2019 (before the global pandemic) more than 20% of employees in Estonian startups were foreign nationals.[33]

The rise of the far right in Estonia (as with much of Europe) has dampened the country's reputation as a welcoming global tech hub in recent years, with the leadership of one party in particular responsible for a series of shocking statements including that "blacks should go back" and deriding former Finnish Prime Minister Sanna Marin as a "salesgirl."[34] According to a 2020 *Politico* article, "recent scandals involving a far-right governing party—under father-and-son duo Mart and Martin Helme—have shaken confidence in the country's leadership, prompting some to question whether Estonia's reputation as a leading tech nation could be at risk."[35] However, the rise of the far right, at least as a political power player, has been significantly tempered in recent years by more mainstream parliamentary coalitions and Estonian society remains one of the most progressive in the region, becoming the first Baltic country to legalize same-sex marriage.[36] Plus, most mainstream Estonian political figures go out of their way to be supportive of the tech industry and not harm it (and Estonia's global reputation), with former President Kaljulaid proudly posting on Twitter about the leadership of Estonia's tech sector.[37]

Unfortunately, geopolitical setbacks outside of the tech industry's control have also hurt the growth of foreign talent and investment coming to Estonia. The war in Ukraine and a looming revanchist Russia has led to fears of the physical safety for potential newcomers and serious risks for investors according to Bolt CEO Markus Villig, who dubbed the additional risk premium Estonia suffers from, "Putin's tax."[38] However, in recent years the country has invested heavily in dual-use technology and "defense-tech" companies, mostly out of necessity and in part because of the larger opportunity to sell defense products and services around the world. The government has been highly supportive of the move, attracting a NATO-backed startup accelerator program[39] to the country and launching a state-backed investment scheme valued at tens of millions of euros.[40]

Estonia's cultural mores also play a role in the development of the startup ecosystem—the country has one of the world's most generous parental leave policies, allowing parents to stay at home with their newborns for multiple years while receiving benefits. Because of its flexibility, it has been held up as a reason why new parents can take time to think about what they want to do next with their lives. Sometimes the answer is to launch a new company, something few new parents elsewhere can seriously contemplate.[41] It is also socially acceptable to take a risk and fail in Estonia, something not true in some other cultures where the stigma of failure can mark someone for life. This is equally true in government where former CIO Siim Sikkut has said that the country "understands that if you try something in earnest and fail—as opposed to failing because of ignorance or misdeeds or whatever—then that's okay. I've never felt in my job that if I failed, something bad would happen.[42]"

The success of Skype and other well-known tech companies has also made tech a prestigious field for the best and brightest in Estonia who choose to create or join new startups rather than going to more traditional fields like finance or academia (and the popularity of their products mean that it is easy enough for young people to explain to their parents what they do for work). On the other hand, the popularity of the tech sector has had a somewhat deleterious impact on the government's ability to recruit the best and brightest since so many prospective civil servants end up as founders and early employees of high-flying tech companies. Although problematic, this is a similar problem faced in countries from Germany to the U.S. where private sector salaries generally outpace those in the public sector, and in the case of the U.S. at least, there is little prestige associated with working in the public sector relative to tech.

Beating all the odds, the country has been wildly successful in developing into a tech hub. It is often listed as a global leader in various rankings of startup ecosystems, like that of venture firm Atomico whose founder Niklas Zennström was also one of the founders of Skype.[43] In 2022, Estonian startups raised more than 1.3 billion euros and contributed 185 million euros in employment taxes.[44] As of early 2024, the country has ten unicorns based in Estonia, developed using Estonian technology, or founded by Estonians,[45] with significant numbers of other local startups moving toward unicorn status. And while the tech scene, like many others around the world, struggles with diversity and gender equity,[46] Startup Estonia and Work in Estonia have made recruiting and integrating foreign and diverse talent to the country a priority in an effort to rectify the situation, facilitating programs like Emerge Estonia, which are meant to help foreign founders better integrate into the local ecosystem.[47] This is also done partly out of demographic necessity given Estonia's status as a low-population country that needs skilled foreign talent to augment its workforce in order to maintain growth.

Today, the tech mafia is far more than just the founders and early employees of Skype and a supportive government—the next generation of startup founders from Wise, Bolt, Pipedrive, and others have become new leaders for the next generation of entrepreneurs, strengthening the Estonian tech mafia. Importantly, not only have these entrepreneurs worked to help other entrepreneurs, but they have also served a critical role in designing the policy and strategy that helped make the country a leading tech hub while serving as evangelists for the country, helping spread the word about Estonia's unique people and culture in tech ecosystems across the world.

At the end of his illustrious career, President Meri gave a speech in which he thanked the Estonian people, "for the beautiful time we have had dragging our cart uphill together."[48] According to President Ilves, who knew Meri well, "His only regret, which he did not express in his speech, but to which he referred in another speech a year earlier, was that Estonia still did not have its own Nokia."[49] President Meri unfortunately passed away in 2006 and did not see the rise of Estonia as a tech hub. However, he would surely be heartened by the fact that his question now had not just one answer but dozens.

10

EXPORTING ESTONIA

Startups like Skype, TransferWise, and Skeleton Technologies have been critical in burnishing Estonia's image as a global tech leader and all-around digital society, creating jobs while serving as a global calling card. However, it has been firms like Cybernetica, an e-government service provider and research center, that have served as the true force multiplier for the government's foreign policy. Because Estonia's political leadership has deeply and intentionally woven its brand in with digital government (or e-government), that branding is seen not just as a tool for attracting investment but also for achieving foreign policy goals. Given its lack of strategic natural resources like oil or cobalt that could be used to achieve geopolitical objectives, the country has turned to companies like Cybernetica to serve as a key export—helping the government forge deep ties with other nations by building their e-government services and selling the expertise garnered from creating a digital nation from scratch.

The story of Cybernetica, which is less a traditional company and more of a phoenix that rose from the ashes of the Soviet Union, is emblematic of the government private-public partnership strategy that has helped lead to a thriving export economy for Estonian e-services that have the added benefit of strengthening diplomatic relationships across the world. The company would eventually be born out of the Institute of Cybernetics of the Academy of Sciences of Estonia, which was founded in the 1960s.[1] Unlike other cybernetics institutes around the Soviet Union at the time that often tackled math and engineering challenges, the Estonian group was focused on computer programming.[2] In the 1960s, the applied research and development laboratory within the institute had already begun work on information security and cryptography,[3] two areas that would later be crucial for the enabling of the Estonian e-state's foundational technologies, digital identity and the X-Road. While it enjoyed a privileged role during the Soviet era as a leading research hub, the institute, like so many other government programs post-independence, faced the challenge of navigating a market economy where resources were scarce. Rather than attempting to stay siloed as an academic institution or government owned firm, the group embraced the market, becoming a private company in 1997,[4] focused on solving some of the most pressing challenges facing the nation, including border security[5] (part of the institute continued to exist under the Tallinn-based TalTech University, but eventually closed in 2016). However, it never lost its deeply academic, research-oriented nature. Even today, Cybernetica maintains an accredited in-house research arm with around fifty researchers (out of around 230 total employees as of early 2024) who do research in the field of cryptography, information security, and beyond.[6]

As one of the few hubs of technical talent in Estonia[7] prior to independence, the group was in a prime position to support the development of the Estonian e-state from the very beginning and many of the academics behind the institute were involved in the early development of the country's first advisory body on ICT.[8] Cybernetica had already experimented with cryptography when it was a research institute, and as the e-state began to take shape, experience with creating secure systems for the transfer of critical information, whether people's tax or health data, became crucial. The company soon became best known for the work of its researchers on the X-Road, one of the foundational pieces of Estonia's e-state.[9]

The X-Road used advanced cryptography to secure user data, keeping it safe from malign actors like the Russian government and various global hacking groups. Cybernetica and its researchers continue to be active developers of the e-state, helping to build an e-customs solution to prepare the country for accession to the EU, developing the i-voting system that allowed Estonians to securely vote from anywhere in the world with an internet connection, and working on the SplitKey technology that supports authentication and digital signatures for e-IDs in the country.[10]

Through its development of e-state solutions for the Estonian government and its applied research, Cybernetica has also become a recognized name and innovator around the world. After developing a virtual air traffic control system in partnership with the Estonian Air Navigation Services to support the latter's goal to digitize regional airports by 2025,[11] Cybernetica decided to spin out the technology, selling it to the publicly traded Adacel, an air traffic management and simulation enterprise.[12] The situation was not an unusual one for the company or for the Estonia tech sector broadly. The government has long prioritized public-private partnerships[13] and, in addition to trying to serve as a proof-of-concept customer whenever possible, sees the commissioning of certain e-services as a "win-win" since once the technology is built and in use in Estonia, it can then be exported, creating local jobs without costing taxpayers.

According to remarks from the longtime e-government advisor and digital champion Linnar Viik in an interview with *The Guardian*,

> "From the early days, government philosophy was not to hire programmers, but to use the services of private companies, which in turn increased the competitiveness of the Estonian IT sector." Case in point: the ID card. "It's private companies who developed and manage the service—and who can now export their new-found competencies to other countries." Viik argues that this benefits both the private sector and the state, who otherwise would not have the resources for ID card technology.[14]

Public-private collaboration has been successful in other jurisdictions as well. According to a 2023 article titled "E-Government Success Stories: Learning from Denmark and Estonia,"

> Collaboration between the public sector and the private sector was also a crucial success element in Denmark and Estonia's

e-government journeys. When Denmark launched NemID, it became available to all Danish banks and e-commerce businesses... Meanwhile, in Estonia, the government collaborated with the private sector to develop the Estonian IT Academy in 2012 to address any potential ICT skills gap. The Estonian government cooperated with local companies such as Cybernetica, Guardtime, and Signwise to deliver the core e-government components.[15]

Cybernetica, as one of the leading developers of e-state solutions since the 1990s, has long been a beneficiary of this strategy.

As of 2023, the company has built e-government systems for more than thirty other countries, including the United States where it has worked in partnership with the Air Force Research Laboratory to develop a cyber threat information sharing system.[16] It has also created or advised on the creation of solutions comparable to Estonia's X-Road, which serves as a digital information exchange, for countries as diverse as Benin and Ukraine,[17] helping to strengthen domestic e-government state capacity while building global ties for the country. And while companies like Cybernetica benefit, many in the government see the private sector as a crucial tool not just for expanding Estonia's brand but for Estonia's security.

President Ilves has called the Estonian data exchange layer "foreign aid on a thumb drive,"[18] but the Estonian e-state has become much more than just a vector for government foreign aid—it allows officials to dramatically extend their reach and influence across the world with no cost to taxpayers, often while making money and creating jobs in Estonia. Today, Estonia's public and private sector collaborators like the e-Governance Academy and e-government service developers like Nortal, Guardtime, and Cybernetica have co-created e-state solutions for Colombia, Tunisia, Ukraine, Benin, Kyrgyzstan, and even Malaysia[19] based on the expertise developed by creating products initially developed for use in Estonia. Public sector officials are occasionally even embedded with private companies as they consult with or build solutions for foreign governments to lend their expertise and, presumably, their credibility.[20] However, doing so creates a risk for the government by associating themselves with specific private companies who they do not control. Having its name attached to a company which commits fraud or even simply fails at a critical e-government project abroad for an allied nation or important trading partner could lead to a serious diplomatic dispute, a massive risk for the Estonian government.

When Estonia exports an e-government product, whether an electronic health record platform or an identity management framework, or simply provides consulting expertise on how a country should go about their digitalization journey, international ties deepen. Given that Estonia can only afford a few dozen diplomatic outposts around the world, with most located in Europe,[21] turning the e-services and expertise that were initially leveraged to make the country more efficient into an export and diplomatic tool enables the country to punch significantly above its weight in foreign affairs. This has become increasingly important given the difficult financial situation facing the country in 2022–2024 thanks to the war in Ukraine, rising energy costs, and high global interest rates. And given the increasingly fraught regional geopolitical situation, with a neighbor that has no compunction in invading the sovereign territory of another nation, it pays for the country to have deep connections with friends across the world who depend on Estonian talent and software to run essential government services.

The closest, albeit imperfect, analogy to how the Estonian government works with e-government companies like Cybernetica, Guardtime, and Nortal is found in the American defense industry. Because of the nature of the items being developed by companies in the defense field, governments create strict export controls on who the products can be sold to and for what uses. No Western government wants an unscrupulous defense contractor selling weapons to a global pariah state or terrorist group (at least, not without permission). Where the analogy fails is that unlike many traditional defense companies, firms like Cybernetica do not create kinetic weapons used in physical combat, and the company also has a thriving commercial product division rather than being solely focused on government contracts.[22]

But governments also often work closely in partnership with defense companies to sell certain products. Defense attachés in embassies around the globe pitch various solutions developed by defense firms to their host nation, often competing for contracts with state-backed or pseudo-independent defense firms from other nations. Governments believe that the sale of military goods will end up benefiting both the domestic defense contractors who will fulfill the order and their own foreign policy, enabling them to deepen defense ties with another nation.

Estonia sees its exports of e-government solutions in much the same way. While Estonia does have a domestic defense industry that has fielded interesting and useful equipment, including a robotics solution fielded in Ukraine much to Russia's consternation,[23] to consider the local defense industry the David to America's, or even Germany's, Goliath would be a dramatic overstatement given its size as of this writing. Instead, since its national security is largely secured by NATO, the country has historically been able to focus on exporting its e-services to accomplish foreign policy objectives. Companies like Cybernetica sell to foreign governments a product used by the Estonian public sector, provide training and expertise, and often embed experts within the bureaucracies of foreign countries, forging long-lasting political and economic ties between the nations, just like the traditional defense industry.

However, one way that Estonia's public-private partnerships have developed is very different relative to America's current defense industry: the revolving door between the private sector and government hasn't led to what could be considered legalized corruption and self-dealing but instead a virtuous cycle of knowledge and culture sharing. This has generated new ideas and policies that have helped maintain Estonia's global tech leadership. In America, the revolving door is much derided, often for good reason. Articles from *The New York Times*,[24] *Politico*,[25] and scores of other publications have shown the deep relationship between former senior defense officials who then take up plum positions at the defense contractors whose products they purchased only a short while before. A 2023 report from the Quincy Institute for Responsible Statecraft showed that twenty-six of the thirty-two four-star officers who retired after June 2018 went to work for the defense industry in some capacity.[26]

EXPORTING ESTONIA

Within Washington, it is an open secret that the way one makes serious money quickly after some years in an underpaid public service position, whether as a staffer in Congress or a career bureaucrat in the Department of Defense, is to work as a lobbyist or directly for a company over which one's former colleagues have regulatory authority or procure large amounts of products or services from, and then to use your relationships and expertise for your new employer's gain. This can be symbiotic in various instances—some officials join companies they see as leaders who they believe deserve to have a larger role in government contracts or advocate for policies they believe are in the best interest of the American people (and their employer)—but often these relationships are predatory and come at the eventual expense of the taxpayer.

In Estonia, people who go through the revolving door between government and the private sector are generally admired for their service. Take Taavi Kotka, who served as Chief Information Officer of the Republic of Estonia from 2013 to 2017 and who received the prestigious official commendation of the White Star III Class Order in 2016 for his work.[27] During his tenure, he helped to create the e-Residency program and the world's first data embassy among other notable achievements. However, before Kotka entered government he was the CEO of Nortal (a portmanteau of Nordic and Talent), a major developer of IT solutions for the Estonian government, including significant work on e-tax and e-health systems.[28]

Instead of concerns over his appointment to the role of CIO of the country because of his links to a private sector contractor which has received a significant number of government contracts (and will likely continue to do so in the future), there was initially excitement about attracting such an esteemed person with relevant skills and knowledge to the public sector. For the government, the use of private sector employees who serve "tours of duty" or go in and out of the government repeatedly has been an important part of facilitating the spread of new ideas and Taavi's situation is not uncommon.

The CEO of Cybernetica, as of this writing, previously served as the director of the Government Strategy Office.[29] Sten Tamkivi, an early Skype employee and tech investor, has had formal and informal advisory roles across various administrations including that of President Ilves.[30] Luukas Ilves, former President Ilves' son, served as CIO of the country from 2022 to 2024 and previously worked at Guardtime, another startup that builds e-government solutions. And people moving between industry and government at every level, not just the top, is common and looked upon as entirely normal. The country has been able to avoid issues with corruption and self-dealing that the revolving door often presents elsewhere because of its unique local and government culture, media environment, and e-government solutions, all of which are difficult for other nations to perfectly emulate. However, there have been exceptions where private sector links raise red flags over concerns both real and imagined, but these cases are few and far between. For cases where concern is warranted, such as an investigation involving a now former Prime Minister,[31] it usually results in a quick firing, robust investigation, and often ostracization as corruption is generally considered unacceptable and a threat to the nation's security.

As a small country, Estonians often like to joke that there aren't six degrees of separation, there's usually one at the most. It's not uncommon for anyone you ask on the street to have some relationship with a well-known politician or public servant: maybe they went to school together, played sports together, or have some extended family link. Almost always citizens know that they can relatively easily reach someone important in the government if they want to. The closeness keeps leaders in the country accountable and humble—nothing brings a politician back to earth faster than receiving a Facebook message from a favorite schoolteacher asking why they have done something controversial.

The local media ecosystem also plays a major role in maintaining accountability. In a close-knit society, there are few secrets that stay secret for long, and the media is extremely free and independent, pulling no punches when reporting on any suspected wrongdoing, a major deterrent for would-be corrupt or self-dealing officials. This applies even to extremely well-regarded politicians such as former Prime Minister Kaja Kallas, who earned the moniker "Iron Lady" for her global leadership in support of Ukraine, and who was put under deep media scrutiny due to her husband's business ties to Russia.[32]

There is also a strong element of duty to country at play. No one thinks that they will go into the government to get rich, and most executive-level leaders like Taavi and Sten serve for free or for minimal pay relative to what they earn in the private sector. But everyone understands the importance of giving back to their country and engaging in public service and are willing to make major personal sacrifices to do so. When asked about his motivation for joining the government, Taavi highlighted the importance of patriotism, saying,

> Part of it is from my upbringing. The thing was that I could not do military service in my youth, but I had been always taught to be a patriot and that you have to do something for your country. The Government CIO role was a chance for me to give back to Estonia.[33]

However, the dramatic rise of the tech sector has seen pay for skilled engineers and executives skyrocket, and the government has not kept up. This has led many to eschew the government for practical reasons, even if they would like to give back to their country, making government service a much deeper sacrifice than it was in the early days of the country. Taavi has made similar remarks. In Siim Sikkut's 2022 book, *Digital Government Excellence*, Taavi emphasized that, for a variety of reasons, "if somebody, a good friend, asked me now if I would suggest working for the government—I would say probably 'no.'"[34] But working in government can still be an opportunity to build unique products and change society in a way one simply can't in the private sector, which serves as a lure for top private sector talent (especially those on garden leave or taking a career break).

The bureaucracy itself also has a unique culture with incentives set up in a way that prioritizes outcomes and talent over seniority. This is in part a legacy of the fact that mostly young people ran the country post re-independence, although the youth influence is waning with time (many of the same young people who took political and government roles after independence remain politically active as of this writing). Those who are top performers can become leaders quickly regardless of age. This also means that there is no "penalty" for leaving the government to do a stint in the private sector and if one brings back new skills that are beneficial for the bureaucracy it is rewarded rather than seen as an affront to a bureaucratic monoculture. Additionally, with a merit-based promotion system, a poorly performing employee after twenty years of public service would be unlikely to be in a senior position, something not shared with the U.S. system which places a premium on seniority-based pay and promotions.

The public sector stays competitive with the private sector in other ways, too (at least relative to governments in the West), such as in basic hiring practices. Even a seemingly small issue like not being able to hire talent fast enough can cause serious problems down the road, as evidenced by issues at both the federal[35] and local[36] levels in the United States, where long wait times lead to unfilled positions, unhappy candidates, and a smaller available talent pool. However, the size of Estonia and its bureaucracy must also be considered as a major factor here; with such a small country and lean bureaucracy, it is much easier to spot low and high performers and act accordingly.

The open architecture of Estonia's e-services and government procurement processes also play a prominent role in mitigating public corruption. Citizens and e-residents can easily create a company online using the e-business registry in a few minutes. However, unlike many countries where business ownership is opaque, the e-business registry is easily searchable for anyone to see who owns a certain company, where it is registered, the number of employees, and even details on their revenue and taxes paid. Journalist Nathan Heller's *New Yorker* profile of the country detailed the setup, saying,

> The openness is startling. Finding the business interests of the rich and powerful—a hefty field of journalism in the United States—takes a moment's research, because every business connection or investment captured in any record in Estonia becomes searchable public information. (An online tool even lets citizens map webs of connection, follow-the-money style.)[37]

Because the Estonian government also makes most procurement information available online, if certain companies owned by former government officials started to net large contracts or anything illicit seemed to be occurring, they would quickly be put under immense public scrutiny. According to a European Commission study on administrative capacity in the European Union, "Estonia's highly developed e-procurement environment and e-procurement portal are frequently referred to as best practice examples for other MS [Member States of the EU] because they are rapid and easy to use."[38] Thanks to Estonia's e-government systems, transparency has helped to eliminate corruption, with the country taking one of the top spots in the nonprofit Transparency.org's Corruption Perceptions Index, which ranks Estonia as less corrupt than many Western European nations like France and Spain, as well as the United States.[39]

Estonia's e-services have served as a domestic tool to improve the delivery of government services but have become much more than another piece of public infrastructure. The country has worked to tie its brand to everything "E" and "digital" whether e-services, digital government, or startups, becoming "e-Estonia" or a "digital nation" as it attempts to shed the unfair and unrealistic post-Soviet image stuck in the minds of many Western audiences and remake itself as a progressive and future-oriented digital society,[40] an image which is well deserved by any account. Those familiar with the nation may be surprised that the country still is viewed in a post-Soviet lens by some. President Ilves has long encountered such views and been quick to educate those who hold them. He recounted in an interview with Stanford Library about having to remind an American delegation from Congress who mentioned needing to teach Central and Eastern Europeans about countering corruption that Estonia and the U.S. had similar levels of corruption at the time.[41] As part of their efforts to become known as a digital society rather than a post-Soviet economy, Estonia has relentlessly leveraged and promoted the private sector and local digital prowess, and engaged in public-private partnerships that could be used to expand the country's reach.

While the relationship between the public and private sectors has been mostly symbiotic and managed to avoid serious corruption issues that plague nations both developed and developing, the strategy has not been without its critics. In an interview with professors Rainer Kattel and Ines Mergel, a former Estonian government official criticized the government's policy of promoting individual companies who develop e-state solutions, saying,

> I have a huge thing with, for example, Cybernetica. I think the government cannot support [only] one company. You have the Estonian government that supported Cybernetica with lobbying, promoting, everything. And they still complain. I hate it. I think it's one of the biggest frauds in Estonia.[42]

As the Estonian private sector continues to grow and the government finds it increasingly difficult if not impossible to promote private sector companies in an equitable manner, such concerns are likely to increase without a change in strategy.

While the number of issues raised with the government's strategy are few and far between, there is no question that promoting specific companies can lead to allegations of favoritism and concerns over corruption and self-dealing. However, because of Estonia's vibrant and open media ecosystem, e-solutions that promote transparency, and the incentive structure in the public sector, the country has avoided major scandal and maintains a status as one of the least corrupt countries in the world. Plus, the strategy to become a digital society and become known globally as "e-Estonia" has paid off handsomely. Estonia is visited by scores of delegations of government officials every year who want to learn its secrets to creating an e-state and the local tech ecosystem has become known globally as a paragon of highly skilled engineering talent, making it easier for Estonian companies to sell their products and fundraise globally.

With the fusion of the public and private sectors and a government strategy that leverages the knowledge and capacity of the latter to achieve the goals of the former in a highly symbiotic relationship, Estonia has managed to accomplish significant foreign policy goals far outstripping what would normally be expected of a country of its size. Through the work of companies like Cybernetica, Nortal, and Guardtime, Estonia has built critical e-government infrastructure for nations in North America, South America, Africa, and the Middle East, forging close diplomatic and economic ties. It did so by serving as a proof-of-concept customer, developing solutions that would later be exported, and cementing its place as a digital government leader with a digital society behind it. But becoming a digital society dependent on technology also meant new risks from old enemies, something the country has learned the hard way.

11

WEB WAR ONE

As Estonians became accustomed to being able to do almost everything online, their increased reliance on the internet created a new vulnerability. The private sector was much the same. Investments in online banking infrastructure and connectivity meant that most people used the internet in both their daily lives and work for critical functions like managing their finances and taking remote meetings. That connectivity meant convenience and gains in productivity, but it also came with new risks. The country had long taken measures to deal with the basic scams, phishing, and online fraud that came along with the web, in part via the implementation of mandatory digital identity. However, few at that time understood the risk of the web becoming a new battlefront.

On 27 April 2007, just three years after joining NATO, Estonia was hit by what may have been the first state-backed cyberattack on another sovereign nation.[1] Former President Ilves referred to it as "Web War One," and the attack made the Estonian people aware of the dangers of the cyber realm before many others around the world had even seriously considered the risks associated with the web.[2] Although there is a lack of definitive proof that Russia was responsible for the attack—one of the benefits of cyberattacks is that their origins are relatively easy to obfuscate—it is generally accepted, at least in Estonia and foreign policy circles,[3] that the attack was either orchestrated by or executed with the tacit agreement of the Russian government. According to Merit Kopli, then editor of local media publication *Postimees*, "The cyber-attacks are from Russia. There is no question. It's political."[4] The digital bombardment came on the back of a seemingly small domestic issue that triggered a massive response from Russia: the moving of a Soviet World War II memorial from its prominent position in downtown Tallinn to a more out-of-the-way location.

The memorial consisted of a bronze statue of a soldier, head slightly bowed with helmet in hand, adjacent to the graves of several Red Army soldiers.[5] It had been installed by Soviet occupation forces as a symbol of the heroism and sacrifice of Soviet soldiers against the Nazis, who had briefly conquered Estonia during World War II. To the Russian-speaking minority in Estonia, it remained a symbol of Soviet (or Russian) heroism and was frequently the site of ceremonies for holidays like Victory Day, which commemorated the triumph over Nazi Germany. However, for the native Estonian population who, while having also been subjugated by Nazi rule, were then brutally occupied by the Soviet Union, the statue was a symbol not of Russian bravery but instead a monument to a harsh decades-long occupation which aroused significant animus.[6]

This natural feeling of hostility, coupled with deteriorating relations with Russia, led to the decision to move the statue. The move caused consternation among the Russian-speaking population, making them feel further marginalized in post-re-independence society. Indeed, Russian speakers in Estonia post-re-independence generally had fewer job prospects and worse health outcomes than Estonian speakers despite any number of efforts undertaken by the government to rectify the situation and provide Estonian-language curricula to the population in the hopes that they would better assimilate into the native Estonian culture. In recent years, Estonia's Internal Security Service (somewhat similar to the FBI in the U.S. or MI5 in the UK) has highlighted the dangers of the diverging health and quality-of-life outcomes. The security service's 2023–2024 annual review of threats highlighted that for some residents, especially those in areas bordering Russia with large Russian-speaking populations, inequality poses a national security risk as,

> A significant number feel that the provision of essential services—education, healthcare, public transport and commerce—is unsatisfactory. This can cause frustration and disappointment among residents, making them more susceptible to manipulation by foreign states or other entities acting in their own interests...[7]

However real the concerns of the Russian-speaking minority, they were further stoked by Kremlin-controlled media outlets. These outlets remained accessible and popular at the time, since most locally produced media was in Estonian and most international media was in English, leading Russian speakers in Estonia to seek out channels with Russian-language content. Thus, about 75% of Russian-speakers in the country watched Russian state media[8] and two of the three most popular channels were produced in Russia.[9] Russian politicians including Putin[10] also aired their grievances directly about the memorial and the Russian government warned of "disastrous" consequences over the plans.[11]

On the day the memorial was to be relocated, a protest at the site turned into a violent riot as the Russian-speaking minority fought the move. Around 1,000 people were arrested with scores injured and one person killed.[12] The physical protests immediately moved online in an orchestrated attack to take down key digital infrastructure. Attacks began on the websites of the parliament, newspapers, banks, even the website for the emergency number.[13] According to reporting from *The Guardian*,

> Vast "botnets"—networks of captured and linked computers—were attempting to bring down computer systems with automated queries as part of a large DDoS (distributed denial-of-service) attack. "Mail-bombing" email barrages and volleys of status and location queries overloaded servers across the country, bringing crucial parts of the Estonian internet to a halt... "War dialling", in which automated phone calls target a company or institution, placed a virtual blockade on phone numbers for government offices and parliament.[14]

The cyber onslaught lasted for three weeks. Though the impact and aggressiveness of the attacks ebbed and flowed, the cumulative effect was serious—especially when it came to international relations and issues related to sovereignty. According to Jaak Aaviksoo, the former Minister of Culture and Education and champion of the Tiger Leap initiative who was serving as Defense Minister during the attack, "This was the first time that a botnet threatened the national security of an entire nation."[15] Thanks to the work of Estonia's cyber-defenders across the private and public sectors, the country was able to mitigate the damage of the cyberattacks on the economy and people's day-to-day lives. Critical was the role of the Computer Emergency Response Team (CERT), which had been formed in Estonia just one year prior and played a critical role in defending against the attack.[16] NATO allies were less able to help in the moment as it was the first time a member state had to defend against a major cyberattack and the alliance was underprepared.[17]

Internationally, the attacks seemed far more devastating than they were in reality for Estonians on the ground, as part of the cyber defenders' strategy to counter the deluge from Russia was to temporarily isolate Estonia's internet from foreign actors. This had the effect of making the Estonian web essentially inaccessible from outside the country.[18] According to journalist Cyrus Farivar's book, *The Internet of Elsewhere*, which recounted the impact of the cyberattack in Estonia,

> In total, more than two hundred websites were targeted. Despite the cyber-damage, many Estonians likely were either unaware of the attacks or didn't seem to care. According to a report in *Postimees* at the time, nearly half of those surveyed were not affected at all.[19]

The direct economic impacts on the nation were estimated to be in the tens of millions of dollars,[20] although some businesses like banks, which attackers had singled out, did expend significant resources countering attacks and lost business due to their services being rendered inoperable.[21] However, the longest length of time a bank's site was down continuously was one and a half hours.[22] In a paper published in the journal *Law, Innovation and Technology*, Dr. Samuli Haataja highlighted how Estonian banks were better prepared than even the government as they had long dealt with the cybercrime prevalent in the region[23] which surely helped in their response to the attack.

The cyberattacks were meant as a show of force from Russia, a real-world demonstration of their capabilities to harm another sovereign nation for taking an action the Kremlin found disagreeable while being able to maintain their supposed innocence, minimizing repercussions from the international community. It became an oft-repeated tactic in both kinetic and non-kinetic contexts. But instead of convincing the Estonian people to kowtow to Russian demands and interference in their domestic affairs, it instead strengthened the resolve of the nation to invest in their own defenses. According to an excerpt about the cyberattack published by the Atlantic Council in 2013, it also led to significant foreign policy changes: "Estonia has become more embedded than ever into Western security and policy institutions, while Russia's cultural and political influence on Estonia has been further reduced."[24]

The shift westward and the increased resolve to strengthen Estonian cyber defenses arrived quickly. The country engaged in new forms of cooperation with NATO and created a Cyber Defence Unit within the Estonian Defence League, a volunteer-based force created in 1918 and re-established in 1991, which was inspired by the American Minutemen who fought the British.[25] The creation of the new unit was spearheaded by then Defense Minister Aaviksoo, building on the already successful concept of reserve forces. President Ilves has compared the new unit with the West's efforts, saying "you've had a weekend-warrior thing where volunteers run off into the woods and do target practice. In Estonia, we have a unit of IT people from banks, software companies who in their spare time for one day a week work on cyber issues."[26] These volunteer cyber warriors could be incorporated into the military chain of command when needed; they also enabled information sharing between private and public sector experts about new threats, and helped to upskill both workers and soldiers at minimal cost.[27] The attack led to an awareness that cyber was something important that affected everyone, but also an area in which every citizen could play a role. The government increased education about digital best practices and made cyber hygiene a priority, requiring ongoing training for government officials and promoting it from the highest levels as the responsibility of every citizen.[28]

In the immediate aftermath of the cyberattack, questions abounded over how NATO should respond and how cyberattacks should be handled in the context of Article V came to the fore as the organization grappled with the proper response. According to Aaviksoo, "At present, NATO does not define cyber-attacks as a clear military action… [T]his matter needs to be resolved in the near future."[29] The German Marshall Fund later reported that while NATO had discussed the cyber threat before,

> after Estonia's digital infrastructure was hit by cyberattacks in 2007, NATO admitted that a confrontation between states might involve a cyber dimension, and at the Bucharest Summit in 2008 adopted its first cyber-defense policy.[30]

However, not until 2016 did then NATO Secretary-General Jens Stoltenberg state that a cyberattack could trigger an Article V response if the impact was serious enough.[31]

Due to their success in causing mass disruption without any real repercussions, Russia has continued to leverage grey-zone or hybrid tactics—malign activities that do not quite rise to the level of outright warfare or that blur the lines of the meaning of warfare like cyberattacks and the weaponization of energy sales—to advance their foreign policy aims, including in Georgia and Ukraine. According to Peter Pomerantsev, author of *Nothing Is True and Everything Is Possible*, a revealing look into modern Russia, "The Russian theory of war allows you to defeat the enemy without ever having to touch him… Estonia was an early experiment in that theory."[32]

However, it was clear that even if NATO was not going to respond directly with a proportional attack of its own, some action would have to be taken, both to show the world its resolve to counter such actions and to prepare member nations for a world in which cyberattacks would increasingly be used by both state and nonstate actors. The response came shortly after the attack with the creation and accreditation of the Cooperative Cyber Defence Centre of Excellence (CCDCOE) in May 2008 in Tallinn. The organization was established to combat cyber threats by supporting information sharing between NATO member states, organizing training exercises, and conducting research on issues related to cyber operations. It had been proposed by Estonia some years prior, but the idea had languished until the 2007 cyberattack spurred several NATO member states to action, functioning as a "wake-up call."[33]

Despite its name, the CCDCOE is not technically a NATO organization. Instead it is accredited by NATO and provides support to member states.[34] The center of excellence is part of a group of more than twenty other centers across NATO that provide support on issues ranging from military medicine to naval mine warfare.[35] The CCDCOE was initially formed by Estonia, Germany, Italy, Latvia, Lithuania, the Slovak Republic, and Spain but has since grown to include more than two dozen other countries including the United States among its "sponsoring nations" and as of 2023 has become the largest NATO center of excellence. In 2023 when celebrating its fifteenth anniversary, the CCDCOE announced that Ireland, Iceland, Japan, and Ukraine had also joined it.[36] Although neither Japan nor Ukraine are NATO treaty signatories, they are both aligned with the alliance.

Bringing Ukraine into the fold of CCDCOE members was indicative of a further geopolitical shift between the West and Russia. Given the unprovoked Russian invasions of Ukraine in 2014 and 2022, it comes as little surprise that the Ukrainian people would seek to strengthen alliances with NATO and the West. While the Ukrainian mainland has become a warzone, so too has Ukrainian cyberspace. Although much-expected major cyberattacks taking down Ukrainian critical infrastructure[37] have yet to occur as of this writing, in large part due to the heroic efforts of Ukrainian government workers and volunteers who have kept infrastructure up and running and also thanks to the support of international partners including U.S. Cyber Command and companies like Microsoft and Amazon Web Services (AWS),[38] the country now continually fends off attacks from Russia. Like Estonia in the early 2000s, Ukraine has become a domain where Russia tests cyber tactics before expanding them to the rest of the world. This is most clearly visible through the release of NotPetya ransomware, which was used to enable a devastating cyberattack which began in 2017 in Ukraine,[39] with the methods and tools behind the attack quickly spreading and used to wreak havoc worldwide, causing billions of dollars in damages.[40]

While some may have taken the impact of NotPetya and the dangers of being dependent on digital to deliver essential government services as a reason to move to a more analog process, the Ukrainians have taken a different approach. In 2014, the country established a state agency for e-governance[41] and began developing e-services like Trembita, a data exchange solution similar to the X-Road, in partnership with Estonia's e-Governance Academy.[42] The focus on e-services dramatically increased after the election of President Zelensky, who had made moving government services online a priority, appointing a top lieutenant to oversee the process.[43] According to a 2024 report by Merle Maigre, a senior cybersecurity expert at the e-Governance Academy who previously served as a director of the CCDCOE,

> In hindsight, Ukrainian officials have been vocal in acknowledging the impact that the cloud migration had on the continuity of core government services and the functioning of the economy. For instance, they claim that "not a single registry has stopped operating" as a result of Russia's cyber offensive.[44]

Like Estonia, because of the real-world expertise garnered by Ukrainian cyber defenders, the country serves as a valuable source of information for the rest of the Western world on how to develop countermeasures and effectively fight a cyber war.

In the CCDCOE's role as a hub in the cyber domain, it often gathers practitioners, policymakers, and other key stakeholders from various governments to conduct research relevant for global audiences and engage in real-world exercises. It runs one of the largest international cyber defense exercises, Locked Shields, which it has held annually since 2010.[45] The 2022 exercise brought together more than 3,000 participants from thirty-eight nations who over the course of four days engaged in active cyber defense exercises.[46] Former Commander of the Estonian Defence Forces, Lieutenant General Johannes Kert, is reported to have remarked, "there are no good solutions when you are under cyber-attack, there are only bad and very bad solutions. All you can do is to train, train and then train some more—to train for the known and educate for the unknown."[47]

The CCDCOE and Estonia have taken this mantra seriously. In part because of the creation of the CCDCOE within its borders and because of the constant threat of another cyberattack, Estonian leaders have continued to prioritize and invest in maintaining a leading role as a cyber hub. The country's cyber leadership was highlighted by President Barack Obama during a 2014 visit to Estonia, who said that, "As a high-tech leader, Estonia is also playing a leading role in protecting NATO from cyber threats."[48] The country continues to be held in extremely high regard for its cyber prowess globally with *Politico* reporting in 2022 that David Cattler, the assistant secretary general for NATO's Joint Intelligence and Security Division, said "It's no accident that the Cyber COE is there given the Estonian government and the Estonian people's investment in high tech and information technology and the related industries."[49] International rankings back up the anecdotal evidence—according to the 2022 Global Cybersecurity Index, the country was ranked third best in the world at cybersecurity.[50]

The investments in cybersecurity made by the country seem to be paying off. In a redux of the events in 2007, the country faced a cyberattack over the moving of another Soviet-era monument in 2022 after Russia's full-scale invasion of Ukraine.[51] However, according to a statement from then CIO of Estonia, Luukas Ilves,[52]

> Yesterday, Estonia was subject to the most extensive cyberattacks it has faced since 2007… The attacks were ineffective. E-Estonia is up and running. Services were not disrupted. With some brief and minor exceptions, websites remained fully available throughout the day. The attack has gone largely unnoticed in Estonia.[53]

The country has also increasingly become a defense-tech hub, with a NATO funded startup accelerator outpost based in Tallinn through the Defence Innovation Accelerator for the North Atlantic (DIANA) initiative, a program meant to seed the development of a new crop of early-stage dual-use technology companies who could provide products necessary for NATO's collective defense.

As of 2023, the associated fund has invested nearly a million euros into local startups.[54]

That is not to say the country is somehow invulnerable. It is a common saying in the cybersecurity world that there are two types of people: those who know they have been hacked and those who don't know it yet. The same applies to nation-states; and, of course, those which are hyper-dependent on digital infrastructure must be hyper-vigilant in protecting against cyber threats. According to Jaan Priisalu, who responded to the 2007 cyberattack while working for Estonia's largest bank, "We live in the future. Online banking, online news, text messages, online shopping—total digitisation has made everything quicker and easier... But it also creates the possibility that we can be thrown back centuries in a couple of seconds."[55] Like Ukraine and Estonia, governments around the world are encountering the dangers of digitalization where a devastating enough cyberattack can plunge society back to the stone age, with a 2024 article from cloud services firm CloudZero estimating the cost of a one-hour internet outage across the U.S. at more than $450 million due to the negative economic impacts associated with the outage.[56] While the thought of such potential economic and social turmoil may cause some to recoil and backpedal towards analog methods, the pace of technological change and the benefits that digitalization bring far outweigh the risks.

12

SOFTWARE EATS GOVERNMENT FOR BREAKFAST

Marc Andreessen, the investor and inventor who would inadvertently play a leading role in the pathway of Estonia's development through his co-creation of the Mosaic browser which helped inspire President Ilves, wrote an op-ed in 2011 that would cause ripples across the tech industry and beyond for years to come.[1] "Why Software Is Eating the World" had a straightforward thesis: the costs to both access the web for consumers and to build internet-based companies were decreasing incredibly rapidly, which meant that the global economy was quickly going to become digital. "Software is eating the world" quickly became a meme in the tech industry and time has proven Andreessen's thesis mostly correct.

Software has reshaped how we shop, how we spend our free time, and even how we meet our romantic partners. Yet, with a few notable exceptions, Western governments have stubbornly resisted this attempted mastication of their services, with only the edges being nibbled at. And to the extent that there has been some attempt to digitize government services, this has tended to be led by private companies rather than public sector institutions. Rather than embrace digital, states have made half-hearted attempts to build solutions or left the development of critical services to the private sector, leading to a "worst of both worlds" scenario as citizens get stuck with lackluster services and states without the internal capacity to build the digital public infrastructure[2] necessary for the current era. Instead of continuing the existing (and clearly failing) strategy of maintaining almost no state capacity—for building or even procuring digital services—government agencies should embrace and invest in digital so that they can either effectively build or buy e-government services.

Given the risks inherent to the digital world, civil servants and politicians must determine whether the benefits of digitalizing critical government operations outweigh the risks. If there was any country that would have a reason to fear those risks, it would be Estonia. By 2007, the country was dependent on the web for a nontrivial portion of its economic output and e-services like digital tax filing and online banking were widely used. Plus, it was still in a geopolitically difficult neighborhood, with an aggressive neighbor who has proven willing to try repeatedly invading physical territory. The 2007 cyberattack showed that Estonia's digital territory was under similar threat and the country had to confront a new vulnerability, this time against warfare that didn't require a physical manifestation to wreak real-world havoc.

The moment represented an opportunity for reflection regarding the long-term strategy of the nation and its focus on tech. Instead of hedging their bets and slowing down or stopping the development of further e-services and investments, the country continued undeterred on its pathway to becoming a digital society. Estonia learned that the only way to address these new risks effectively was to focus on digital technologies and build the capacity needed to combat the threats that come with all new technologies, whether the internet or AI. Besides, once a country goes digital, going back would be akin to deciding to return to the stone age after the introduction of the Model T because the stone age posed fewer dangers from car accidents. President Ilves, who lived in California for several years while teaching at Stanford University after his presidency, was struck by the backwardness of local government services when registering his child for school:

SOFTWARE EATS GOVERNMENT FOR BREAKFAST

> The only thing that was in this process that was a nod to the modern era was the Xerox machine, but otherwise all this is identical to how life was led in the 1950s... so then you have this paradox. You are living in the mecca of technology doing amazing things, but your ordinary life as a citizen... it's a paper world. A paper world with serial processing, the way bureaucracy has been run ever since it was invented.[3]

For Estonia, rather than being cowed by the cyberattack and going back to a paper world, the government and private sector saw it as a learning opportunity to strengthen the nation's defenses—a lesson that the rest of the world would do well to learn. Surviving "Web War One" soon became part of the country's brand and a badge of honor, contributing to its then near mythical digital status. Soon, entities like the e-Governance Academy and the CCDCOE would be teaching the rest of the world about cyber risks and how to fend off would-be attackers. The country decided to continue its journey of becoming a digital powerhouse, where most government services could be accessed online and where society was equipped with the skills needed to succeed in a digital world.

The conscious decision to focus on building up e-services, paired with the clear need for the public and private sector to work together to tackle serious threats like major cyberattacks, created a mentality among the general population that prioritized public service and a mentality within the state that prioritized attracting the best and brightest. This meant that when it came to matters of state capacity on digital issues, the Estonian ecosystem of officials, informal advisors, and strategically aligned vendors were well placed to make good decisions when designing future e-services.

Most Western countries have taken a very different approach. In places like the U.S. and the UK, building state capacity when it came to digital, whether the capacity needed to build technology internally or effectively procure and outsource complex IT solutions, wasn't a priority until recently. Instead, major e-government efforts—from Healthcare.gov to repeated failed efforts to modernize Veterans Affairs IT systems[4]—were ineffectively outsourced without any clear strategy or project management guiding the development of such digital public infrastructure.

Of course, there have been powerful constituencies who stood to gain from inefficient and massive outsourcing who pushed for the public sector to function as a checkbook rather than a builder or competent project manager. Professor Mariana Mazzucato and Rosie Collington's 2023 book, *The Big Con: How the Consulting Industry Weakens Our Businesses, Infantilizes Our Governments and Warps Our Economies*, highlights the risks of outsourcing, saying, "The more governments and businesses outsource, the less they know how to do."[5] This creates a self-fulfilling cycle as a weakened public sector depends on private sector firms with their own interests and misaligned incentives.

The situation is much the same elsewhere in the world. In 2024, Britain was rocked by the Post Office Horizon scandal, which Professor John Naughton said was, "the product of a flawed and sometimes clueless IT procurement system to which the British state has for decades been addicted. The system eventually procured... was a sprawling, computer bug-filled monster."[6] The Horizon software was meant to modernize Britain's Post Office but instead was so full of bugs that it became a plague on the postal system, leading to hundreds of workers being accused, tried, and sometimes convicted of theft erroneously due to IT system errors, even leading to some taking their own lives.[7] Despite various bugs being known since 1998[8] and multiple convictions being overturned,[9] the situation only received attention from the highest levels of government in early 2024 after a popular television show aired, leading to mass outcry and a public response from the Prime Minister.[10]

The reasons why the Horizon Post Office Scandal took place and lasted as long as it did are numerous, but there is a mounting consensus that one of the culprits is the lack of digital knowledge within the British government leading to a flawed procurement contract for an IT system that they didn't understand. Alice Moore, an assistant professor at the University of Birmingham, authored a piece in which she claimed, "The Post Office didn't have the expertise it needed to understand what was going wrong..."[11]

SOFTWARE EATS GOVERNMENT FOR BREAKFAST

In comparison, while Estonia has long outsourced significant portions of its digital government infrastructure, it has done so in a strategic manner that prioritized maintaining enough internal state capacity to adequately manage vendors and projects, rather than hollowing out all technology-related state capacity. In addition, such outsourcing contracts were frequently fulfilled by local companies whose employees freely flowed between the government and private sector, minimizing losses in competence while maintaining deep links between the public and private sectors. Thanks to their strong state capacity (among other factors), Estonia's e-government development has been mostly unmarred by controversy or significant cost overruns.

In 2021, the state budgeted about 190 million euros for IT related expenses,[12] with a significant portion of the amount going to maintenance and cybersecurity,[13] and in 2022 the government allocated an additional 30 million euros for cybersecurity given increased threat levels.[14] Part of the reason for Estonia's success is that the government has also consistently funded digital government projects with clear budgets over long periods of time, with about 1% of national expenditures allocated for the development of e-services and related costs.[15] This consistency has helped enable long-term planning and stability, with hugely beneficial impacts for the e-government ecosystem. However, this is not always the case, as highlighted by Kaire Kasearu, Head of IT Financing at the Ministry of Economic Affairs and Communications and Ott Velsberg, Government Chief Data Officer in a 2021 article on Estonia's e-government failures. According to Kasearu, "Politicians are keen to promote a specific solution and argue for securing investments for it but often downplay or simply forget altogether the necessity to keep the system running after it has been implemented."[16]

Estonia's development of e-government infrastructure stands out as having few catastrophic failures despite a culture of intense experimentation. This stands in especially stark contrast with the United States where most government IT projects fail, with some estimates putting the success rate at only 13% for large projects,[17] despite the U.S. government spending more than $100 billion on information technology every year.[18] According to the former CTO of the Environmental Protection Agency Greg Godbout,

> the federal government still struggles to buy, build, and operate technology in a speedy, modern, scalable way... While there have been success stories, like the IRS's direct file tool and electronic passport renewal, most government technology and delivery practices remain antiquated and the replacement process remains too slow.[19]

This is not to say that the Estonian government is totally immune to failure. In 2022, the country acknowledged a humiliating setback in the development of the Schengen Information System, a pan-EU project managed by the European Commission when then Minister of the Interior Lauri Läänemets was forced to apologize for not being able to complete the project on time.[20]

However, due to the robust digital expertise within the state, many more potential IT failures are prevented. With experts who can build systems or have the knowledge to effectively procure from the private sector, projects are more effectively scoped, developed, and rolled out. According to Aare Lapõnin, who helped develop some of the earliest e-government systems in Estonia,

> In my view, the one extremely important pillar that was present in Estonia was a certain state of mind of many technical people, who were close enough to decision-making, had the technical skills and the "let's do it" attitude, were able to spot opportunities for meaningful public administration improvement projects, and were able to execute those projects.[21]

As software continues to eat the world, it's only a matter of time before it comes for more government services. This is especially true in the United States where so many services that are considered "public" elsewhere are privately administered, from electronic health records to tax filing. Private companies with strong profit motives have already embraced the web as a cost-saving measure and simply because it is what consumers have come to expect. Estonia has managed to avoid the major risks of digitalization because of its willingness to invest in domestic capacity within the government and creating a culture of trust within the civil service which allows the internal development of e-services or intelligent outsourcing under the watchful eye of public servants who know how e-government infrastructure should be built. That culture and capacity has yet to be developed in much of the West, to the detriment of the public. For countries to be successful in a digital world, they must embrace digital as wholeheartedly as Estonia. Half-measures will inevitably lead to more failure and disappointment while countries who leverage technology pull further and further ahead.

There are scores of reasons why Estonia has been able to digitalize its society effectively and leapfrog from a poor post-Soviet economy to a global leader in e-government and tech—existential threats to the nation which forced cohesion and action, resource constraints that necessitated innovative thinking and action, bold leadership from young technically literate leaders, even occasional bits of pure luck like being physically close enough to receive television and radio signals from Finland during the Soviet occupation and gaining independence at the dawn of the modern web. But no reason is more important than the unique culture that undergirds all of Estonian society. Famed management consultant Peter Drucker coined the unforgettable line, "Culture eats strategy for breakfast."[22] There is no exception for government. Estonia shows that building the right culture (and capacity) is critical for any nation that hopes to succeed in the digital era and while building a new organizational culture is difficult, it is by no means impossible.

PART II

LESSONS LEARNED

Human Resources > Natural Resources: Early on in President Ilves' tenure as a government official he realized that all successful nations have one thing in common—a highly skilled citizenry. For a country to survive, especially one like Estonia with few natural advantages, it would require galvanizing the entire nation into a workforce that could thrive in the newly ordained information age. As in Lee Kuan Yew's Singapore, superior intelligence, discipline, and ingenuity would be a substitute for resources. This meant schools had to be connected, children taught digital skills, and entrepreneurship promoted so that Estonia could compete on the global stage. But it wasn't just the private sector workforce that needed the right skills, the public sector needed them too, and working to ensure that the public sector also had access to the best talent has been a core part of Estonia's successful modernization. There is also a side benefit that efficiency can enable "less state" and deflate bloated bureaucracies through process and technology improvements.

Put a Leash on It (The Private Sector): As a country forced into communism, Estonia understood better than most the dangers of the state having too much power and control. However, it also understood that private sector incentives aren't always aligned with national interests and in certain cases the state would have to engage in oversight and work to create a symbiotic relationship. This is especially true when it comes to private providers of products and services that are core to the traditional role of government, from defense to taxation. The great Charlie Munger, cofounder of Berkshire Hathaway, once said, "Show me the incentive and I'll show you the outcome." This holds just as true for government employees as it does for private sector consultancies.

Software Is Eating Government, Slowly: Despite the risks inherent to digitalization and the difficulty many governments face in transitioning from analog to digital services, the drumbeat of technological progress never ceases and governments will have little choice but to modernize. Those that hope to do it well and avoid being left behind, vulnerable to cyberattacks and predatory consultancies, should instead embrace the shift and work to build digital capacity. Like Estonia in the early 1990s, there is also an opportunity for early adopters and first movers to leapfrog competitors by adopting new technologies—think of how much a country could save by harnessing AI to create precision medicines for its population that would drive down healthcare costs.

There Is No E-Panacea: In an inherently physical world, there will be many services that require real government workers to deliver them. No matter how digital a country, not every government position can be automated. State capacity cannot be limited to digital—if a government runs a healthcare system it will need adequate numbers of doctors and specialists to serve the population, regardless of whether they see the doctor in person or via telemedicine. This means that governments have to learn how to cost-effectively deliver their services in both analog and digital form, especially given social mores towards things like voting in the West.

PART III

THE FUTURE OF THE E-STATE

13

INFORMATION AGE

The e-Estonia Briefing Center is the central nervous system of the government's efforts to educate global audiences about the country's unique history, digitalization journey, and the people behind its rise as a global e-government and tech leader. Based in a modern office park close to the Tallinn airport, the outside of the building is unassuming. But once inside, visitors are transported to something more akin to the headquarters of a Silicon Valley startup than what one would expect from a government office. For the thousands who make the pilgrimage every year to the tech and e-government mecca that is Estonia, it rarely disappoints. But despite the numerous delegations[1] sent by countries like the U.S. and Germany and the fact that former President Ilves has called it "one of the most important institutions in Estonia,"[2] few of those visiting have truly absorbed the lessons of Estonia and applied them in a meaningful way. This is unfortunate, because in a rapidly digitalizing world (and one where advances in technology like AI promise to further increase the pace of change) Estonia serves as an important lesson for how a nation can adapt to the digital era in a way that is both democratic and effective.

There are many reasons why governments like the U.S., Germany, and Japan have failed to learn lessons from Estonia's journey, but their central problem is that they face the Innovator's

Dilemma: victims of their own earlier successes, they are trapped in a legacy mentality, holding on to what they already have rather than embracing the risks that come with innovation and disruption. In short, they have much to lose and by trying to hang on to what they have, they may end up losing everything. These countries were hugely successful during the industrial age when mass and manufacturing capacity ruled. But while they are still industrial giants (in relative terms, despite the rise of China) and in some cases home to private sector tech giants, America, Germany, and Japan's government bureaucracies predate the information revolution of the 1990s and their processes generally reflect it. Although some countries were early adopters of basic computing technologies—the U.S. federal government was the largest customer of the U.S. computer industry in the 1960s[3]—their bureaucracies, except in limited instances, have not adapted to the information age.

Government operations remain stuck in the past despite advances in computing and artificial intelligence which could and should facilitate dramatic changes to how the government functions. Instead of reforming, bureaucracies have been encumbered with scores of new rules and regulations that make it near impossible to change the system. According to former U.S. Deputy CTO Jennifer Pahlka,

> When we speak of "legacy systems" in government, it does not mean simply that they are old. It means that we are grappling with the legacy of decades of competing interests, power struggles, creative work-arounds, and make-dos that are opportune at the time but unmanageable in the long run.[4]

Political scientist Steven Teles popularized the term kludgeocracy in his seminal essay on the topic to describe the increasing complexity and incoherence of American government operations which led to excessive costs and failures thanks to the addition of "inelegant patches put in place to solve an unexpected problem."[5] This legacy of workarounds and quick fixes adds up over time and leads to a kludgeocracy where the state is rendered slower, more inefficient, and less competitive over time thanks to a continuous buildup of "kludge."

INFORMATION AGE

Former President Ilves has commented that laws, like culture, are the software of society.[6] As such, the legal system can be just as encumbered by "legacy software" in the form of "kludge" as any IT system, with outcomes just as deleterious. Take federal hiring—many well-intentioned laws and rules to prioritize fairness, equity, and marginalized groups or those who have served their country have led to a situation which makes it incredibly difficult to hire quickly—necessitating legal intervention to enable workarounds to hire for specific roles (often for IT and scientific professionals where government salaries are noncompetitive). In the United States, the legal system has become a drag on the bureaucracy despite the desperately hard work of government intrapreneurs like Pahlka as well as experts like Marina Nitze and Nick Sinai whose 2023 book *Hack Your Bureaucracy* shared lessons and tactics they developed together while working in the Obama administration on government modernization efforts. The book was hailed as "A master class on intrapreneurship" by former Google CEO Eric Schmidt thanks to its practical and forthright examples[7] but the fact that such a book and the lessons within it were relevant to those navigating the federal bureaucracy serve as a damning indictment as to the state of the government's overall modernization efforts. As tech pioneer Steve Blank said in a 2024 article titled "Why Innovation Heroes Are a Sign of a Dysfunctional Organization," "Innovation heroics are a symptom of a lack of an innovation doctrine."[8] The need for constant heroics within U.S. government agencies is clearly indicative of a lack of a clear doctrine for innovation and a sign of organizational dysfunction.

The work of other government intrapreneurs and groups like the U.S. Digital Service (which has been renamed the United Sates DOGE Service under the leadership of Elon Musk) and the U.S. Government's Technology Transformation Service's 18F division in supporting agencies like the Centers for Disease Control and Prevention on pandemic preparedness tech has been heroic and has saved taxpayers billions,[9] but not yet led to meaningful lasting changes across the federal bureaucracy. The bureaucracy still suffers from a stultifying mentality that attempts to avoid risk wherever possible, often hiding behind the law to avoid change along with the risk and uncertainty which come with it. According to Pahlka, the state of the American bureaucracy is in crisis and things clearly must change, in part to show that the democratic system can still deliver for the average American who is frequently frustrated by bureaucracy.[10]

For a long time, Estonia's successes and much of the West's failures when it came to e-government services wouldn't have been all that important in the grand scheme of things. Germany, Japan, America, and scores of others are still industrial-age powerhouses and although their bureaucracies are stuck in the twentieth century, they are successful nations for the most part. The United States has had, as of this writing, an exceptional showing in economic growth post-pandemic[11] and it would be a leap to consider Germany or Japan anywhere close to failed states. However, the information age,[12] in which power is shifting to those who have mastered the digital world, has only truly just begun. The forces that underpin what defines a nation's strength are shifting rapidly. To be a powerful state prior to the information age, one needed access to natural resources as well as a large industrial base, a strong military, and the workforce to go along with it. There's a reason why America was long called the "arsenal" of democracy given its enormous manufacturing base and defense industry (although it has voluntarily ceded that role thanks to massive industrial offshoring, a sclerotic defense sector, and poor policies).

Today, the forces that propelled America, Germany, and Japan to global leadership are changing. America is no longer the world's sole manufacturing superpower as it was post-World War II, and its global influence is on the wane. Small armed groups can increasingly and seemingly easily disrupt global trade as evidenced by the activities of the Houthis in Yemen. Software is becoming more important in global affairs, whether used for cyberattacks, autonomous weapons systems, or simply increased economic efficiency. And according to former Armenian President Sarkissian, small states have increasing amounts of power over global affairs relative to their industrial heavyweight counterparts, since,

> A small, tech-savvy state can now remotely sabotage and paralyse large states. Technology has eroded the capacity of large powers

to remain the predominant centres of progress and achievement. Where once large powers shaped the fate of the world, today small nations are endowed with the prowess to compete with them.[13]

The sometimes prescient (and often controversial) 1997 book, *The Sovereign Individual: Mastering the Transition to the Information Age* by Lord William Rees-Mogg and James Dale Davidson posited that the information age would lead to dramatic changes in who held power. It would, suggested the authors, potentially result in the destruction of the traditional nation-state as the dominant locus of sovereignty, hastening the shift from industrial-sized mass production to small groups coming together to create value on the web. It even predicted the rising power of wealthy and hyper-mobile individuals who they called "Sovereign Individuals." According to an updated preface for the book by famed (or infamous) investor and political donor Peter Thiel,

> Those trends—winner-take-all economics, jurisdictional competition, the shift away from mass production, and the arguable obsolescence of interstate warfare—are still at work today. The rise of China is less a refutation of Rees-Mogg and Davidson than a dramatic raising of the stakes they described.[14]

In a 2024 article, Vox discussed the possibility that the world is entering a "neomedieval" period which, according to a RAND Corporation study the article details, is "characterized by weakening states, fragmenting societies, imbalanced economies, pervasive threats, and the informalization of warfare."[15] While the pen has long been mightier than the sword and communication has enabled dramatic geopolitical changes, the rapidity of the proliferation of connected devices means that potentially anyone with a smartphone can change the balance of power. It is unlikely that the industrial-era nation-state model will survive long without some adjustment. Steve Jurvetson, whose role as a venture capitalist often requires him to consider what the future will look like, has posited that, "a nation state that doesn't embrace technology in every aspect, let's say with AI, with genetically modified organisms, with nuclear energy will fall behind pretty quickly," emphasizing that while this has been true for decades it is now accelerating even more rapidly.[16]

A critical feature of the information age made clear in *The Sovereign Individual* is that tremendous amounts of wealth can be created by exceedingly small groups of people with minimal physical assets or ties. Skype was both Estonia's first unicorn startup and a perfect example of the importance of the then nascent digital economy. Between the early founders and engineers, the Skype team was able to develop a product that scaled to millions of users globally in a few years and was a company that could be run anywhere there was an internet connection and the workers with the skills to build it. Indeed, the fact that it was built by Swedish, Danish, and Estonian entrepreneurs shows the power of the nascent web to counter traditional positive agglomeration effects by enabling a distributed yet highly efficient workforce. The downstream effects of Skype's success would be powerful for the Estonian nation, leading to a cascade of new investment and innovation. Without it, it is unclear whether the Estonian state and startup sector would have been able to get where it is today, or at least as quickly as it did.

Less than a decade after eBay's acquisition of Skype, Facebook (now Meta) purchased WhatsApp for a whopping $19 billion in 2014.[17] The company had more than 450 million users of its messaging platform across the world,[18] a staggering rise for a product launched five years prior. It was especially popular outside the U.S. where data was cheaper than text messaging, for which mobile carriers often still charged a per message fee.[19] Even more incredible than the acquisition price and growth of the firm was the fact that a company with hundreds of millions of users had just over fifty employees at the time of the acquisition.[20] At a $19 billion price tag, that equated to about $350 million dollars of value generated per employee. While WhatsApp was launched in the Bay Area, it was created by Ukrainian immigrant Jan Koum along with his colleague Brian Acton[21] and those fifty-five employees could have just as easily been physically located in Tallinn or Texas.

With the rapid development and adoption of AI-powered tools post launch of OpenAI's ChatGPT, from copilots for software development and writing to automated image generation, it is possible that soon even smaller groups of people will be able to create dramatically more value with the help of artificially intelligent agents acting on their behalf. Already, various investors and AI leaders, including OpenAI CEO Sam Altman, are predicting that a single individual with the help of AI can create a unicorn tech company.[22] Creative industries may be equally affected with tech investor Zak Kukoff predicting that new text-to-video AI models will enable tiny teams to create a box office hit in the next five years.[23]

When a few people can create tremendous amounts of value they can become more mobile, picking and choosing their preferred government jurisdictions. This spells potential danger for countries dependent on taxes from the wealthy such as tech moguls purchasing or attempting to purchase foreign citizenship in New Zealand,[24] Indonesia,[25] and Cyprus.[26] This trend, along with the growth of the golden visa industry,[27] highlights the increased mobility of the wealthy and the continued unbundling of the "services" traditionally provided by a single nation. According to reporting by CNBC in 2024, "Henley & Partners, a law firm that specializes in high-net-worth citizenships, said Americans now outnumber every other nationality when it comes to securing alternative residences or added citizenships."[28]

The COVID-19 pandemic supercharged the number of remote workers[29] (often also referred to as digital nomads, although there are subtle distinctions with the latter being defined by their frequent global travel whereas remote workers are often stationary but do not go into a traditional corporate office). According to the workforce consultancy MBO partners, the total number of American digital nomads was more than 17 million as of 2023.[30] No longer tied to a single location for their work thanks to tools like Zoom, GitHub, and Slack along with increased global connectivity, they can much more easily shop for their ideal jurisdiction, one which provides better government services or delivers on basics like affordable housing. Case in point: in just the first nine months of 2022 temporary residence permits for Americans to live in Mexico increased by 85% from 2019 as Americans vote with their feet and head south of the border (it remains to be seen how many such workers relocate permanently).[31]

Sensing the revenue opportunity from attracting wealthy foreigners who are already employed and won't displace local workers, countries from Georgia to Barbados have responded by crafting bespoke visas. These visas often promise concierge-level services or provide generous tax breaks,[32] with the Barbados Welcome Stamp initiative enabling remote workers who stay in the country for twelve months to pay no local income tax.[33] According to reporting by the *Financial Times*, "Rough estimates by the IMF in 2022 found that increased remote working reallocates about $40bn of the income tax that workers pay globally."[34] While the article frames the risk of losing such taxes as small, tens of billions of dollars up for grabs is a massive opportunity for a small nation like Estonia, and indeed the article goes on to say that IMF research found that "Small emerging market economies 'with below-average tax rates and good remote work capability' typically gain the most from the trend."[35] However, it is not yet clear whether the rise in digital nomad visas and international remote work is an aberration in a rapidly fracturing world where globalization and even international travel may be increasingly hindered by geopolitical factors and where major corporations seek to exert more control and supervision over employee's lives, as exemplified by Amazon's 2024 mandate for all U.S. employees to return to the office five days a week.[36]

With potentially billions of dollars up for grabs, sub-national jurisdictions are naturally trying to get in on the action as well. The Tulsa Remote program provides a $10,000 grant and three years of free coworking space (among other perks) for those willing to relocate to the city. According to research conducted by Harvard in 2024, the program brought in millions of dollars in additional income and sales taxes to the community from the new remote workers.[37] This sub-national poaching of talent may become dramatically more popular given the ease of moving to a new state or city compared to a new country (even with bespoke visas and relocation programs), pushing the fiscal effects of increased talent mobility down to state and local coffers rather than at the national level. Talent competition may also lead to positive regulatory changes at the local and state levels thanks to jurisdictions competing for talent, whether in tax policy, zoning reform to drive down housing prices, or differentiated social policies on topics not regulated at the federal level.

Unsurprisingly, Estonia was at the forefront of the movement to attract global talent, having begun preparing a visa aimed specifically at digital nomads in 2018 (in part based on the urging of local entrepreneur Karoli Hindriks), one of the first such programs in the world.[38] According to Hindriks in a 2018 interview with the publication *Quartz*,

> In terms of the future of work we are all navigating, there is no policy to support the new ways of working... A digital nomad visa represents a breakthrough in the way governments support today's mobile workforce.[39]

However, the country suffered the bad luck of officially launching the program in August of 2020 during the heart of the COVID-19 pandemic[40] and faces a delicate balancing act with a far-right constituency unfriendly to most forms of immigration (an issue not uncommon in the rest of the EU). It also had to compete with significantly more tourist-friendly jurisdictions who had quickly launched similar visas to counter job and tax losses from the steep decline in short-stay tourism to their countries in the wake of the early economic devastation from the pandemic. Despite this, as of early 2024, more than 500 digital nomad visas have been issued, with around 50 million euros added to the economy as a result,[41] and the government has estimated that more than 50,000 digital nomads from around the world visited Estonia in 2023 alone.[42] With a newly materialized market millions-strong with tremendous spending power, sometimes to the point of distorting local markets and causing complaints of gentrification[43] from locals who find themselves priced out of housing in budding digital nomad hubs, there is little wonder that countries would respond to the opportunity.

The idea that wealth can be created by fewer, more mobile individuals with increased optionality is important not just because of the potential impact to the tax base of various jurisdictions, whether national or local, but because of other downstream effects on national power and the global order. Imagine the unfortunately increasingly plausible scenario of Russia invading Estonia. Even barring a NATO response, much of Estonia's top-tier tech talent (who are generally highly mobile and can work anywhere with an internet connection) could easily flee to neighboring countries. Those so inclined could then wage a covert cyberwar against Russia, wreaking havoc on banking and other critical connected infrastructure from behind a keyboard. Russia could capture physical territory and assets like Estonia's well-placed port in Tallinn, but almost surely at a devastating cost.

Many of the high-skilled residents would flee and set up shop once again in tech and expat havens like Toronto or Silicon Valley, or perhaps even follow in the steps of the people of Kiribati who purchased twenty square kilometers of land on a Fijian island for several million dollars to ensure some level of continuity for the physical nation-state in case of environmental catastrophe.[44] This would deny Russia some of Estonia's most productive "assets" and those remaining, even if forced to collaborate and integrate into the Russian tech sector, would be more likely to become saboteurs and impediments to Russia's economy than contributors.

At the same time, the nature of warfare is undergoing a shift that will likely benefit small, nimble, wealthy groups. Instead of large infantry and mechanized units (the Russian war against Ukraine being a notable and continually evolving exception), the world is much more likely in future conflicts to see autonomous air-, land-, and sea-based drone swarms, intelligent loitering munitions, and AI-powered cyberweapons used by digitally proficient nations, especially as birth rates in developed nations decrease and the value of human life increases. Former Google CEO Eric Schmidt, who according to *Bloomberg* "has for years acted as a liaison between Silicon Valley and the US government" as former Chair of the influential National Security Commission on Artificial Intelligence and founder of the Special Competitive Studies Project, has taken an even stronger view than many about the future of warfare, telling an audience at the Future Investment Initiative in Saudi Arabia in late 2024 that the U.S. should trade its tanks for drones, saying "Give them away. Buy a drone instead."[45] Schmidt has put his money (and time) where his mouth is, founding a defense-oriented drone company in 2023 which happens to be incorporated in Estonia according to reporting by Forbes.[46] Some of these robotic and AI-enabled weapons may be controlled locally or remotely by the actual citizens of the nation, some may be directed by private military contractors from around the world, loyal to the highest bidder. And while attempts to use overwhelming manpower may become a feature of future warfare as it has been during the conflict in Ukraine, the rise of mercenary outfits like the Wagner Group and Russia's recruitment of foreign soldiers show that a large standing domestic armed force may not be necessary to field the personnel needed to fight a war. In fact, it may be preferable for wealthy countries who could simply pay soldiers from developing nations to fight on their behalf rather than risk their own citizens.

If this seems like science fiction, consider that during several recent American administrations, the U.S. waged multiple wars in the Middle East with minimal military casualties enabled in large part by leveraging increasingly automated and remotely managed systems in the form of drones (once America had gained air supremacy)[47] and by contracting out huge parts of the war effort to private defense firms like Blackwater and Halliburton.[48] Given the renewed interest in the defense tech sector in light of the war in Ukraine and increased tensions between great powers,[49] the game-changing nature of Turkey's Bayraktar drones in the conflict over Nagorno Karabakh,[50] and the continued importance of private military contractors in places like the Sahel,[51] these trends seem poised to continue. However, changes are likely to be uneven and highly context-dependent. Admiral James Stavridis, former Supreme Allied Commander of NATO and co-author of *2034: A Novel of the Next World War*, has compared the changing dynamics between newer and more traditional forms of warfare to a dimmer switch where militaries must effectively balance investment in new methodologies and technologies like low-cost drones versus expensive aircraft carriers rather than simply flipping an on-off switch between one and the other.[52]

Unlike larger global counterparts, Estonia has adapted to become a leader in the information age. It is small and nimble with a highly educated and entrepreneurial workforce, plus a strong local culture and cohesive national identity. But most importantly, it has incorporated the early lessons from companies like Skype into its national strategy, running the country much more like a startup than other nations, working to deliver products and services that are useful, reliable, and cost-effective for their end users (also known as customers). While politicians in the United States often bloviate about running the country more like a business to the point it has become a groan-inducing cliché, there is something to the comparison—like a company, a country is expected to deliver a product or service to its customers, whether an individual citizen or a local corporation. One can argue exactly what to include in those products and services and who the actual customers should be, but on almost every level, formerly dominant industrial-age nations are failing. According to a Huffington Post article, in 2021, the U.S. Social Security Administration asked disability benefit recipients for feedback on one of their processes for retaining benefits. In a dramatic understatement, the report said that "This process can be extremely onerous," before adding "One recipient told the agency that it was, 'More frightening than having cancer—twice.'"[53]

According to the OECD, the average single worker in America paid 30.5% of their income in taxes in 2022[54] and a 2022 report from the U.S. Chamber of Commerce found that 10.5 billion hours were spent on government paper forms in 2021 with an effective cost of $117 billion.[55] In a country like Estonia, the paper processes could easily just have been automated and benefits provided proactively to those in need, saving time and money for both citizen and state. The potential savings from following Estonia's lead in e-government are not insignificant—McKinsey has estimated that government digitization just at current technology levels could generate a trillion dollars in new global growth[56] and a 2016 *Harvard Business Review* article estimated that bureaucracy costs America over $3 trillion in lost economic output every year.[57]

Despite the urgent need and best efforts of numerous hardworking government intrapreneurs, many government modernization projects are failures,[58] leaving the U.S. extremely far from Estonia's e-government example and highlighting just how unequipped the country is for the information age. In 2023, Congressman Ro Khanna, who represents much of Silicon Valley and has previously introduced legislation on modernizing government websites and services,[59] voiced his concerns over the pace of government service modernization, saying,

> I think it's just a sense of frustration that constituents express that why can't the government deliver better and get it delivered more efficiently? Why is it that government websites are such a mess? Why is it that I feel like I'm 20 years back when I visit a government website?[60]

All this means that not only are Western nations primed for failure in a connected world, but even the democratic foundations of Western society are at risk. The public has exceedingly low trust in the government and their elected officials, and no wonder when nearly every project is delivered late and over budget (when delivered at all). When people are giving 30% of their income to the state only to see their hard-earned dollars wasted on failed consulting contracts and deals that look suspiciously like kickbacks to government consulting and defense firms whose boards and executive ranks are populated by former government officials, it is little wonder that more people are willing to vote for someone willing to burn the system down or who pledges to get things done no matter the cost (or simply move to another country). Rather than doubling down on a failing strategy, the West can turn to Estonia, which has found a way to thrive in an increasingly digital world by operating less like a kludge-filled bureaucracy and more like a startup.

14

COUNTRY-AS-A-SERVICE

Most industrial-era powers like the U.S. and Germany (and even some of the stodgier departments of the Estonian government) have yet to acknowledge that they run a type of business and rarely think of the average citizen as their "customer." While countries may not have the same profit motive that companies do, they certainly are increasingly experiencing competitive market pressures for top talent and are expected to deliver services efficiently and effectively to their customers. In a 2024 *Financial Times* interview about countries' efforts to court digital nomads and remote workers, a KPMG global mobility tax expert emphasized that, "The driving force behind digital nomad visas is that these countries are in competition with each other over labour."[1]

This goes to show that like any business, if there is a competitor who can provide better services at a better value, customers are eventually liable to move. According to Taavi Kotka, who served as CIO of Estonia,

> As in the private sector, the public sector has the responsibility of retaining citizens and adding new residents, businesses, and income generators to the roster to improve the economic condition. Global competition among states is growing, however, and doing the same-old, same-old won't equal results.[2]

Thanks to Estonia's investments in digital government infrastructure, workforce education, connectivity, and state capacity, the country has positioned itself as a global leader of the digital age.

Early Skype investor Tim Draper is highly aligned with Kotka's views, saying in a 2016 interview where he lauded Estonia's strategic foresight that, "I believe the governments are now in competition with one another... but for us. We are now their customers; people, entrepreneurs, money, business—all of it."[3] The analogy of country-as-company is imperfect: moving between nations involves dramatically more friction than switching from one business to another and those jurisdictions that consider themselves of particularly high value put up plenty of barriers for would-be customers (sometimes in the most literal sense).

However, the rise of computing and the dramatic increases in wealth seen post-industrial revolution mean that there are both new modalities for migration and people with the education and resources to take advantage of them. While much focus is on large-scale human migration given its economic and sociopolitical impacts, there are other government services that today create significant government revenues and which may become increasingly competitive, such as business domiciliation (the state of Delaware derived about a quarter of its annual budget in 2022 from business incorporation fees[4] and the British Virgin Islands generate more than 50% of their government revenues from business incorporation fees[5]).

For example, in just a few moments, an entrepreneur based in Lagos can use Stripe Atlas (a service run by the startup Stripe which found a business opportunity due to market inefficiencies created in part by poor government digital infrastructure) to create a Delaware-based corporation entirely online. Moments later, they can easily raise funding from global investors and serve clients from anywhere in the world, entirely outside of Nigeria's corporate jurisdiction. In fact, as of 2021 Nigeria was one of Stripe Atlas' largest growing customer bases with 400% year-over-year growth,[6] and in early 2024 one in six new Delaware corporations was incorporated using the product.[7] According to a Stripe survey of more than 1,000 Atlas customers,

COUNTRY-AS-A-SERVICE

When asked what their greatest barrier to entrepreneurship was, a plurality of founders (43%) said, "bureaucracy". Would-be entrepreneurs with great ideas have told us that they've abandoned starting a company because of process roadblocks or lengthy response times, like receiving a tax ID number over fax machine. We believe forming and running a company should be efficient end-to-end, and governments can digitize or optimize services to help entrepreneurs meet regulatory requirements, as Estonia has been doing with their e-Residency program.[8]

And access to the web is rapidly expanding—as of 2023 there were more than 5.4 billion mobile subscribers around the world[9] and companies like Elon Musk's Starlink are rapidly increasing access to low-cost, high-speed internet through low-earth orbit satellite constellations. While this may bring increased pressures on countries like Nigeria, the potential losses for a country like the U.S. are far greater given the much greater distance there is to plunge.

Because of the shift to digital and advances in technology more broadly, one could easily believe that the value derived from citizenship or residency in an industrial-era nation has been on the wane for years. This is especially true for the United States, at least in relative terms given how powerful the country was decades ago and how far it could fall. From factors like the value of an EU or Singaporean passport outstripping America's,[10] difficulty in accessing essential services like quality healthcare or housing, poor public safety, and minimal public transportation infrastructure, America's value proposition is increasingly weak. Thanks to a failure to invest in technological state capacity and poorly planned policies, the bureaucracy isn't providing adequate digital services, while making the services that are provided increasingly irrelevant or of poor quality—basic cable in the age of streaming services.

Realistically, the whole of the American government bureaucracy responsible for delivering essential services is failing to keep up with the modern era. In a world where a car can be ordered to almost any pinpoint on a screen in moments and autonomous vehicles increasingly roam city streets, having to physically go to the Department of Motor Vehicles with a utility bill, lease, and a driver's license or passport just to register a vehicle seems decidedly archaic. As does filing a tax return through a third-party provider for a fee when the government has all the data needed to instead tell you directly what you still owe (at least, for most taxpayers). Add to the list having to bring physical copies of your health records to your doctor when you change physicians, or worse having to needlessly repeat expensive tests, rather than sharing access to your e-health profile and full medical history with a keystroke. According to David Eaves, Associate Professor in Digital Government at UCL, the impact of COVID-19 highlighted the importance of public digital services as,

> They are critical infrastructure for well-functioning societies in the 21st century. Governments that were able to verify citizens' identities online; securely and instantaneously provide them cash; and safely exchange their information across agencies were able to deliver on the promise of a 21st-century safety net. Those that could not, generally did not.[11]

All of this adds up to a not-so-simple calculus for the value a nation provides its "customers." The problem that the United States faces, along with other industrial-era leaders like France or Germany, is that the value they provide is rapidly declining as they fail to keep up with new technological trends that are reshaping global economic forces and failing to create useful digital infrastructure (physical infrastructure being another issue, albeit one well beyond the scope of this book). At the same time, it is getting easier to unbundle government services thanks to near-ubiquitous connectivity with cheap, reliable computers and smartphones.

Take the example of the Nigerian entrepreneur starting a business in Delaware via Stripe Atlas—this is enabled almost entirely thanks to advances in technology and decreasing costs for connectivity. More value-added activities beyond business incorporation are likely to follow as enterprising individuals leverage the web and technology to earn and save money by circumventing local market inefficiencies and barriers. Some countries may fight by putting up legal roadblocks that stop activities like business incorporation in foreign jurisdictions or even implementing significant tax penalties on those who wish to "exit" their jurisdiction and move to another. However, others will act more like a business and compete for a piece of these newly developing markets rather than putting up barriers.

Estonia has decided on the business-oriented approach, taking it far further than most other nations by not just attempting to adapt to the forces roiling global power structures but by embracing them and positioning the country as a global leader of the digital age. It has done so by working to modernize the country's government services and creating a workforce trained for a digital world and by maintaining its customer-centric mindset focused on delivering valuable products for their "customers."

Estonia has also created a new level of customer-centric government product design with a concept developed by Taavi Kotka during his tenure as government CIO and co-founder of the e-Residency program. Coined "Country-as-a-Service,"[12] a play on the term used for the common software delivery business model, software-as-a-service (SaaS), the idea of the Country-as-a-Service mindset is elegant in its simplicity: unbundle or develop scalable government services in a way that they can be used not just by Estonian citizens but by those from any nation. This concept built on the already pervasive mentality in the domestic public sector that services should be useful and solve real problems but extended it to potential customers around the world who might leverage Estonian government services in some way—in essence it meant that Estonia would go out and compete for "customers" globally against foreign nations who previously may have believed they had a monopoly over their citizens.

The Country-as-a-Service concept is remarkably similar to some of the ideas espoused in *The Sovereign Individual* more than a decade earlier, although whether a prescient prediction from the authors, a self-fulfilling prophecy, or a coincidence is unclear. The ability for the Estonian government to easily execute such a strategy was enabled by the existing digital identity system and e-services built over prior decades in the country. Once the state could serve the Estonian diaspora, it led to a new line of thinking as to what the nation could feasibly undertake, with Taavi remarking,

> If it is possible to offer a convenient and effective e-services environment to expatriate Estonians, why not also offer it to non-

Estonians, even those who do not reside in Estonia, who need better everyday solutions than those offered by their own states?[13]

However, according to Taavi, the concept has thus far failed to take hold across the entire bureaucracy, with some senior leaders expressing skepticism as to how things like police forces could be considered a "service."[14] And while in a globalized world pre-COVID and before Russia's full-scale invasion of Ukraine such ideas may have seemed possible (if not highly controversial), geopolitical changes and rising nationalism may mean that such efforts will face tremendous pushback and lawfare from other nations who may feel threatened. Regardless, while Taavi's concept faces serious headwinds, including within Estonia as entrenched actors push back against change they see as unnecessary and often unhelpful, it is a concept worth exploring. This is especially true as wealthy entrepreneurs like Balaji Srinivasan, former CTO of Coinbase and general partner at Andreessen Horowitz, and Peter Thiel fund various startups and projects that seek to compete with the existing nation-state through companies like Prospera in Honduras[15] (which was dealt a blow in 2024 when the Honduran Supreme Court declared the scheme enabling Prospera unconstitutional)[16] and Praxis[17] which may lead to an increasingly competitive dynamic for citizen-customers and their state-related expenditures. Some investors and entrepreneurs have even more radical ideas as to the future of the nation-state with Tim Draper questioning the need for physical countries with borders at all thanks to the success of cryptocurrencies and alternative government services.[18]

If Taavi's Country-as-a-Service concept does take root (or if people like Draper or Thiel's predictions come to pass) soon an American-born citizen might live part of the year in Porto and part in Tulsa while operating a business domiciled in Estonia with a retirement plan managed by the Netherlands and a healthcare policy administered via the Thai government underwritten by Lloyd's of London—with major ramifications for government revenues and service delivery. Absorbing the breadth of the emerging challenge facing traditional industrial democracies is critical: the way businesses are built, people live, and how they choose and consume government services (and pay taxes) are changing rapidly

COUNTRY-AS-A-SERVICE

as technological progress enables increasingly mobility and empowers individuals and small groups. Only by similarly beginning to approach the delivery of public goods as a true "service" that the government provides and thinking of people as customers can governments hope to compete. And the best way to understand the potential of the Country-as-a-Service concept (whether it becomes a global norm or ends up a flash in the pan) is through e-Residency, a program that has profoundly shaped Estonia's recent history.

15

TEN MILLION ESTONIANS

His Holiness Pope Francis,[1] former chancellor of Germany Angela Merkel,[2] and President of France Emmanuel Macron[3] have at least one thing in common—they've all been e-residents of Estonia. The e-Residency program was the Estonian government's first serious foray into providing a product built specifically for non-Estonians, becoming the flagship "Country-as-a-Service" product.[4] The program was hailed as "The Future of Immigration"[5] by Vice and quickly become one of the key calling cards of the Estonian state. As of early 2024, the program boasts more than 100,000 e-residents from 176 countries, who contribute millions of euros of tax revenue annually[6] and who have a direct connection to Estonia and the country's e-state thanks to the program. Despite those laudable achievements the program has failed to be as revolutionary as initially expected and the underlying value proposition has deteriorated as other jurisdictions create rival programs.

Although the concept that would become e-Residency had been floating around for some time, the program truly got underway with a pitch by Taavi Kotka, Siim Sikkut, and Ruth Annus at an event put on by the Estonian Development Fund and entrepreneur and Sten Tamkivi in the spring of 2014[7] to garner new ideas and build on Estonia's momentum as an emerging digital leader. The winner would receive both public and private sector support for their initiative. Taavi was then serving as CIO of the nation after a storied career in the private sector. Siim was a longtime public servant with a hand in many of the most ambitious government innovation projects, including working to make Estonia a hub for autonomous vehicles and the renewal of the national digital health agenda. Ruth joined the Ministry of the Interior in 2001 and quickly established herself as an expert in citizenship and migration policy among other fields. Together, the group was well positioned to develop the concept that would become the e-Residency program based on a simple premise: If the e-government infrastructure already developed in Estonia, from digital identity to the X-Road, was easily scalable well past the 1.3 million Estonians at home and abroad already using it, then it could also be used by third-country nationals. Even better, it would essentially cost nothing to the state because they were using pre-existing tools and services.[8] The project they pitched was titled "10 million e-Estonians by 2025," with the intention of building a program that would "make Estonia great: make sure that at least 10 million people around the world choose to associate with Estonia via e-identities."[9]

The initial product was simple: an e-ID, without a photo so it wouldn't be mistaken for a travel document (although thanks to the program name "e-Residency," there would frequently be confusion over whether the program granted physical residency benefits), issued by the Estonian government to applicants who paid a fee and passed a background check. Just like any Estonian citizen's e-ID, it would give the cardholder access to the country's digital infrastructure. However, exactly what e-residents would use that access for was somewhat of a mystery at the time despite the lofty pitch.

Even with the lack of clarity over the exact benefit of the program for e-residents, the problem it was intended to tackle was one critical to the future of the Estonian nation. By the 2010s the country had solved many pressing economic and defense challenges and was now a member of the European Union and NATO. However, the country was encountering a new danger in the form of a looming expected decline in the size of the working population.[10] For a nation of 1.3 million with a difficult-to-master language and unique culture, the decline was potentially existential. According to Andrus Viirg, who headed Enterprise Estonia's Silicon Valley outpost for more than a decade, population growth was one of the main challenges facing the country.[11] A declining population meant both economic slowdown and the potential erasure of the Estonian culture. While the traditional response to this challenge is to incentivize local population growth or spur inward migration, the country is not a top immigration destination thanks to its isolation, harsh climate, and complex language, and at times more people were leaving the country than moving there[12] (although in recent years the country has had positive rates of immigration which made up for a negative birth rate).[13] According to Taavi, "Nobody wants to come to Estonia physically."[14]

That's where the e-Residency program came in. It would allow people around the world to connect with the country, learn about its unique culture, and leverage its digital infrastructure, bringing in virtual residents rather than physical ones. While e-residents wouldn't counter a low birth rate among Estonians, it would add "population" to the country's economy, helping maintain growth and spread its culture worldwide. It was also posited that the program would have national security benefits through a form of geoeconomic statecraft. Even though e-residents wouldn't be there to pick up physical arms in the event of an invasion, they could rally in support of the Estonian cause. As Singapore's Lee Kuan Yew had once put it, small states must "seek a maximum number of friends."[15]

The idea was a winner, not only in the contest but with politicians from almost every party, who soon provided significant support. That included rapidly passing tailor-made legislation in October 2014, only a few months after the initial pitch that allowed the issuing of e-ID to foreign nationals living abroad, enabling the small team to launch a pilot version of the program in December of the same year.[16] The expected use case by early adopters remained vague, even to the founders of the program, but the outpouring of public support was tremendous. According to Taavi, "We wanted to run this like a government startup... and with any startup, you [continue to] develop your product when you get to market."[17] Within the first three months of the program being announced publicly, thousands of people from around the world expressed interest and wanted to find a way to be involved despite a dearth of specifics on what exactly e-Residency was or the benefits it granted. So many people went to the program's landing page in the first twenty-four hours that the servers crashed.[18] Even though the founding team had only started with the general idea of using Estonia's digital infrastructure to enlarge the nation, it was clear that there was a nontrivial segment of the global population eager for any sort of government innovation and it was worthwhile to continue developing the product.

The first three e-resident digital identity cards were issued that December to Edward Lucas, Steve Jurvetson, and Tim Draper. Lucas had worked in Estonia before the fall of the Soviet Union as a journalist and founder of the first English-language weekly in the Baltic states[19] and has been described as someone with,

> a habit of popping up at pivotal moments in European history. In March 1990, shortly after Lithuania declared independence from the Soviet Union, the Economist editor caught a flight to Vilnius and received the first Lithuanian visa: number 0001, a stamp-sized chink in the Iron Curtain that got him arrested and deported by Soviet authorities.[20]

Several decades later, he became e-resident number one, but his document was granted to him by his friend and then President of Estonia, Toomas Hendrik Ilves,[21] and fortunately did not result in detention or deportation. The honor of becoming e-residents number two and three respectively would be granted to Steve Jurvetson and to Tim Draper, two of the earliest investors in Skype[22] via the venture firm Draper, Fisher, Jurvetson.[23]

Their new e-Residency e-IDs gave Lucas, Jurvetson, and Draper access to Estonia's digital infrastructure. At this early stage in the program, author and reporter Cyrus Farivar commented that, "it's not immediately clear what notable, obvious, and meaningful benefits being a virtual Estonian would bring," although he also noted that he intended to apply to try it out.[24] Farivar's questions were eminently reasonable. For Estonian citizens, digital identity serves as the key to online access for most public services, from filing taxes to voting and using digital signatures. But, other than as an interesting collector's item or conversation piece, it wasn't clear how e-Residency would benefit foreign nationals. However, there were early signs of where the program could provide value to more global citizens like Edward Lucas, who said at the program launch that he was tempted to use his e-Residency card to facilitate running his company, which he used to manage his freelance work while he was employed at *The Economist* in London.[25] At the same time, nearly half of the early applicants to the program stated that they would also be interested in creating an Estonian company.[26]

Within the first several years of the program's existence it became apparent to the team—which had expanded to include then Managing Director Kaspar Korjus and Ott Vatter (who would later become Managing Director of the program and known as someone who had a knack for building team spirit, who put his friends and country first, and who always had the perfect recommendation for any occasion before his untimely passing in 2024) along with a host of Estonian and foreign experts based in a previously abandoned paper mill[27]—that there were three core user groups of e-Residency. The first was made up of foreigners (mainly from neighboring countries like Russia and Finland) who already had a business in Estonia they managed remotely and who would find being able to do things digitally a major time saver. The second was made up of Estoniaphiles who had some sort of connection with the country or wanted to build one and saw the e-Residency program as an avenue to create a tangible tie to the nation. Those ties could quickly become meaningful, as made clear in a *Quartz* article by an e-resident who claimed that, despite being an American citizen, she felt most loyal to Estonia as an early e-resident of the country.[28] The third constituency was made up of entrepreneurs from around the globe who saw e-Residency as an easy way to access Estonia's well-established business environment where taxes were low and bureaucracy was minimal (although there was often overlap between the latter two groups as many in the former would end up starting businesses and many in the latter would become fans of the country). To these entrepreneurs, the value proposition was obvious. Not only did they get access to the business and legal infrastructure of a transparent and safe economy in the European Union, but they could do it at minimal cost.

The impact of being able to gain access to a safe and low-cost business environment where rule of law was respected was profound for entrepreneurs around the world. In 2017, e-Residency partnered with the United Nations Conference on Trade and Development to support women in Delhi, India, with the creation of location-independent businesses that they could run from their homes, helping them create financial independence and access global markets.[29] Instability in foreign jurisdictions due to geopolitical concerns has also frequently made Estonia a potential safe harbor. For example, in the immediate aftermath of their vote to leave the European Union, the e-Residency program saw a tenfold increase in applications from the UK.[30]

Surprisingly, even to the founders of the program, a large subsegment of e-Residency business owners emerged, mainly in the form of Western European nationals who are frustrated with bureaucracy and byzantine business environments with arcane rules. For the governments of Germany, Spain, and France whose own citizens have created thousands of businesses in Estonia, the program's success should be a wakeup call. In a 2024 interview with *Forbes*, e-Residency Managing Director Liina Vahtras talked about the reasons why EU nationals joined the program, mentioning that "One German entrepreneur said to me, you have no idea of how lucky you are. It's so easy to start a business."[31] According to Tim Draper, e-Residency could serve as a competitive force for government, saying:

> It [e-Residency] makes it so that, if one country is not performing as well as another country, people are going to the one that is performing better... We're about to go into a very interesting time where a lot of governments can become virtual.[32]

However, with the number of EU citizens joining the program, Estonia risks blowback from countries who may see e-Residency more as a government-enabled tax avoidance scheme than the future of global commerce, even though such tax avoidance thus far remains a theoretical issue. This is an especially fraught risk for the small nation which depends largely on the collective goodwill of allied foreign nations for its defense against malign foreign powers.

Though the program suffered repeated growing pains, including having to figure out how to get every Estonian embassy around the world to issue e-Residency cards for new members in person (somewhat ironically, the "digital nation" requires physical interaction to join the program, although in late 2024 e-Residency's managing director announced a call for tenders for a mobile data collection solution with the intent of providing a digital alternative to the current physical e-Residency card by 2027).[33] The program also had to engage in complex negotiations with local banks who were hesitant to provide accounts for non-residents and risk running afoul of global financial services regulators, but still, it grew steadily. Within ten years, the program contributed more than 200 million euros to state coffers[34] and became a tangible symbol of the Estonian nation around the world. By 2023, the program had already welcomed more than 90,000 e-residents, a virtual population size larger than Tartu, the second-largest city in the country.[35] It also led to an outpouring of global media coverage and support, with thousands of mentions in global outlets[36] viewed by millions, many of whom never previously associated Estonia with innovation and e-government and digital leadership. According to Sten Tamkivi, who became an advocate for the program from the moment it was pitched and served on the organization's board,[37] "The level of media coverage for e-Residency, and by association for digital Estonia, is immeasurably higher than the actual revenue it generates."[38]

The program also yielded any number of opportunities for building global economic and diplomatic relationships. The government granted e-Residency status to a range of notable businesspeople and public officials with whom it hoped to deepen ties, like venture capitalist and author Guy Kawasaki and investor Ben Horowitz.[39] According to Siim Sikkut,

> E-Residency has been noticed in diplomatic or cultural circles. It has caused excitement, made Estonia look better. While decisions of whether to station NATO troops here are likely not made based

on e-Residency, it supports things like that indirectly... Estonia's e-residents make up a community interested in Estonia, our circle of friends in the world.[40]

The program has even inspired other countries from Azerbaijan[41] to Lithuania[42] and Ukraine,[43] to invest more heavily in e-services, often leaning on Estonian expertise to build their systems.

Despite the success of the program, it has been the target of criticism and suspicion for harms both real and imagined. The 2023 paper, "Online Incorporation Platforms in Estonia and Beyond: How Administrative Spillover Effects Hamper International Taxation," documents multiple concerns. The issues include administrative challenges related to the rapid implementation of a novel program designed by just a few people without broader consultation of government stakeholders,[44] a cumbersome process for monitoring e-resident companies,[45] and challenges with obtaining information on potential e-residents for the purposes of conducting background checks.[46] There have also been historical issues with bad actors using the program for fraudulent purposes, including the creation of a fake university[47] and crypto scams,[48] with the Council of Europe's anti-money laundering committee expressing concerns over the risk mitigation measures the program is taking to prevent misuse of the financial system.[49]

From the beginning of the program, the founders were aware of the concerns other nations would have over the possible facilitation of tax avoidance, a realistic issue given that as of December 2023 40% of e-Residency company founders were EU nationals[50] who often hailed from high-tax and high-bureaucracy jurisdictions. In 2016, Taavi said in a Vice op-ed about e-Residency that global entrepreneurs "look to other countries, not because they are looking for a tax haven, but because they have been prevented from incorporating and maintaining a business, due to barriers from their own government,"[51] with the op-ed further specifying where non-resident Estonian business owners should pay taxes.[52]

Despite concerns over financial chicanery, the most damning criticism of the program comes not because of its perceived risks, but because of its perceived lack of ambition. According to an interview with former Prime Minister Mart Laar, who had been both advocate and instigator for many of the most ambitious efforts undertaken in the country's modern history, he didn't believe that e-Residency was a crazy enough idea.[53] This can also be seen in the creeping minimization of the ambitions of the program. While it was initially pitched as a program to bring 10 million foreigners into the fold of the Estonian e-state (with the added benefit of showing that the state could also create disruptive products competitive with the private sector),[54] the program has now focused mainly on driving increased tax revenue through business incorporation. Even a decade later, this remains its sole use case for e-residents and the program risks increased global competition from copycat programs and even from private sector companies like Stripe Atlas.

Despite the initial pitch, which was relatively radical at least for the time, Siim has mentioned in subsequent interviews about the program when asked about the goal of 10 million e-Estonians that "Only journalists keep track of that target,"[55] and Taavi has said that the goal was simply a made-up number to serve as a rallying cry to get people excited in the initiative.[56] Now well past the startup phase, the product has evolved relatively little since its introduction. Ambitious potential new ideas like the ill-fated Estcoin,[57] a proposal to explore developing a cryptocurrency token and launching an initial coin offering via the e-Residency program that was immediately panned by the European Central Bank[58] due to its potential for undermining the Euro, have fallen by the wayside. However, the idea did help to keep Estonia in the news with then Managing Director Kaspar Korjus estimating that media coverage of the outlandish proposal was seen by around 200 million people worldwide.[59] And long gone are the heady days when intriguing questions like whether e-residents could be given voting rights or representation in the European Union were seriously considered[60] thanks to bureaucratic malaise and a feeling that the program is doing "well enough" without engaging in controversial or disruptive activities. External events have also served as a factor, with the rise of the far-right making such ideas more controversial domestically and the war in Ukraine shifting government focus towards defense along with more tangible concerns like energy prices and inflation.

But despite its flaws and lack of ambition, the e-Residency has undoubtedly been a net gain for Estonia. It has created scores of local jobs, hundreds of millions of euros in tax revenue[61] and less tangible benefits, helping to put Estonia on the proverbial map as a global leader in e-services and as a top business-friendly jurisdiction. More than that, its success helped to prove the validity of the Country-as-a-Service concept, showing that as the world is increasingly connected, governments can (and almost certainly will) compete with one another to provide services for global citizens. It has also spawned similar programs across the world hoping to mirror the Estonian program's success, including Azerbaijan's m-residency which was created in large part thanks to the support of an Estonian consultancy,[62] showing the global attraction of the idea (as well as how prolific the Estonian e-governance solution export business is).

But despite protestations by some affiliated with the program that e-Residency isn't a replacement for citizenship or national identity, many who have encountered it have walked away wondering if there isn't a broader lesson about the future of the nation-state to be found within the e-Residency program and the Country-as-a-Service model. As journalist Nathan Heller posited,

> Today, the old fatuities of the nation-state are showing signs of crisis. Formerly imperialist powers have withered into nationalism (as in Brexit) and separatism (Scotland, Catalonia). New powers, such as the Islamic State, have redefined nationhood by ideological acculturation. It is possible to imagine a future in which nationality is determined not so much by where you live as by what you log on to.[63]

The e-Residency program may yet be the harbinger of a new form of statehood, one based on where you log on rather than where you reside. But perhaps more importantly, the move to the web has also demonstrated that for countries like Ukraine that have come under physical attack or those that very well may like Estonia, their states have a potential lifeline in the digital world.

16

CLOUD COUNTRY

On 24 February 2022, Ukraine was invaded once again by Russia in a brutal attempt to impose Vladimir Putin's will on the country, which was forging a path towards deeper integration with the West. Much of the media's focus in the early days of the invasion was naturally on the heroic fighting by Ukrainians and the world's shock at such a brazen act of terror taking place in Europe. While Ukrainian troops fought bravely on the frontlines, a similarly audacious scramble to protect Ukraine's digital sovereignty was unfolding mostly outside the media's gaze. Thanks to the digitalization of many Ukrainian government services, significant numbers of records and e-government services were online. This would serve as both a lifeline for Ukraine in the days and months after the invasion as the country worked to provide services to a people at war, but it also created a cyber vulnerability that had to be addressed.

Part of President Zelensky's pitch to the Ukrainian people was a pledge to modernize the government bureaucracy by creating the "state in a smartphone," and to bring all public services online by 2025.[1] This would not only make life easier for the average person but help to eliminate the corruption that had plagued the nation in much the same way that Estonia's early investments in e-government infrastructure stymied public sector corruption.[2]

REBOOTING A NATION

Far from an idle political promise, soon after taking office Zelensky created a Ministry of Digital Transformation tasked with delivering on his ambitious vision. To run it, he empowered a cadre of young IT professionals, including entrepreneur Mykhailo Fedorov, who would serve as both Minister of Digital Transformation and Deputy Prime Minister in the administration, telegraphing to all how important the task of digital transformation was to the Zelensky administration.[3] According to Jaanika Merilo, an Estonian who lived and worked in Ukraine for a great deal of her career, including a stint serving as an advisor to the Ukrainian Minister of Digital Affairs,[4] "it was a political decision that the e-government is a priority. The ministry was established and now over three hundred people work in [the] Ministry of Digital Transformation."[5] By bringing in and empowering young people who often were only in their twenties and thirties, Ukraine's efforts resembled those of 1990s Estonia when young ambitious people accomplished heroic feats against all odds.[6]

In the years prior to the invasion, the ministry's efforts were a resounding success. The state quickly launched the Diia application as the basis of the state in the smartphone. According to Fedorov, this is "the foundation stone of Ukraine's emerging digital ecosystem," and "was a big step towards the 'paperless' vision that lies at the core of the Ministry of Digital Transformation philosophy."[7] The ministry also created an online portal that allowed for virtual business registration, enabling Ukraine to build a program that could compete with the Estonian e-Residency program; it eventually launched in a closed beta test in December 2023[8] and was implemented with the support of the Estonian e-Governance Academy.[9] Even more impressive were the rates of user adoption for Diia. By 2021, the application counted more than a quarter of the Ukrainian population as users and scores of public services from a digital identity solution to unemployment assistance were made accessible online.[10] And as of September 2024, Diia even lets couples get married online, namely for those separated by the war.[11] Today, Ukraine is increasingly acknowledged as a global e-government leader thanks to its development of the Diia platform, which was ranked as one of the best inventions of 2024 by

Time,[12] as well as the nation's work in the cyber realm defending the country from Russian attacks. Thanks to its expertise, Ukraine's advice is increasingly sought out by countries seeking to digitize. In 2024, the Ministry of Digital Transformation signed a memorandum with the nation of Ecuador to assess their digital infrastructure and provide recommendations and support.[13]

However, like Estonia's foray into e-government years earlier, expanding state services to the digital world came with risks from cyber threats which could wipe out e-government infrastructure with the press of a key (or kinetic attacks that could take out a government-run data center). Ukraine had long been a target of Russian cyberattacks and was no stranger to the dangers posed by state and nonstate hackers targeting critical infrastructure, having dealt with attacks on government services, the electrical grid, and the banking sector.[14] What the government had likely not expected was an attack which threatened the data centers holding information critical to running state operations and the personal data of millions of Ukrainians from historical real estate transactions to business ownership. Existing laws on the books were focused on data sovereignty and maintaining such data within the country's physical borders, rather than preparing for a conflict-fueled future which threatened the country's territorial integrity.[15]

In the days leading up to the invasion the likelihood of a major attack and its devastating impact became clearer. The Ukrainian government quickly came to understand the necessity of having backups of their data elsewhere, where an errant artillery shell (or targeted missile) striking a physical data center wouldn't lead to the loss of information critical to the day-to-day functioning of society. The parliament quickly passed legislation enabling data to be moved to the cloud and hosted physically outside the country[16] and requested help given the urgency of the situation. Amazon's Web Services division, AWS, responded by working with the government to send Snowballs—ruggedized data storage devices that look like a cross between a suitcase and a 1990s era home desktop—to quickly transfer as much critical data as possible out of the country after hashing out which data to prioritize on pen and paper over lunch in the Ukrainian Embassy in London.[17]

The timely efforts of AWS and the Ministry of Digital Transformation were fruitful and more than 10 million gigabytes of government and economic data was successfully moved outside of the country to an Amazon facility for safekeeping and uploading to the cloud.[18] This occurred just before a cruise missile targeted a government data center and Russian cyberattacks tried to destroy the information held on other physical servers in the country.[19] According to Liam Maxwell, the former Chief Technology Officer and Technology Adviser for the UK government who later became Government Transformation Director at AWS, waiting for the Snowballs at the baggage claim was tense: "Here's government in a box, literally."[20] The stakes were extremely high, especially in the early days of the invasion when it was unclear how much territory Russia would to take. Minister Fedorov would later say, "AWS made one of the biggest contributions to Ukraine's victory by providing the Ukrainian government with access and resources for migrating to the cloud and securing critical information."[21] The company was awarded the Ukraine peace prize a few months later for its work with Minister Fedorov saying that "Amazon AWS literally saved our digital infrastructure.[22]

AWS itself said that,

> While preserving Ukrainian data is an important driver of this massive effort, perhaps more important is how this data is being used to help Ukraine's people. For example, migrating information about educational degrees and university curricula from the Ministry of Education and Science helps thousands of students prove the validity of their education when applying for work or pursuing another degree. The effort also helps students continue with their final exams at the end of this school year, whether they are in Poland, Moldova, or Ukraine.[23]

Fortunately for Ukraine and the West, the country was able to mostly withstand Russia's repeated offensives and, as of this writing, has fought to a near standstill, preserving much of the country's territorial integrity. At the same time, the country has managed not only to maintain its digital services but to continually expand them.

Since the onset of the war, the ministry has rolled out new digital products including providing financial assistance to those affected by the war via the Diia app, with nearly 5 million applications for funds and hundreds of millions of dollars disbursed[24] as well as the creation of a unique application to help fight back against the Russian invasion. The eVorog chatbot allowed vetted civilians to report on troop movements and provide intelligence to the Ukrainian military; it has played a significant role in the war effort, helping Ukrainian armed forces and intelligence agencies effectively combat Russian attacks and plan devastating counter attacks while shepherding in a new era of warfare where an everyday person equipped with a smartphone can make a nontrivial contribution to the war effort.[25] An *Economist* report on the tool compared it to World War II, when "British volunteers phoned in early warning of air raids. Ukraine's digitally enabled iteration is more powerful. It makes it more expensive for Russia to occupy Ukrainian territory and gives Ukrainians in occupied areas a way to resist."[26]

Despite (or perhaps because of) the war, the Digital Transformation Ministry ceaselessly innovated and adapted, not only safeguarding citizen data through the cloud but growing the number of users of the Diia application to 63% of the population, according to a survey in late 2022.[27] This includes Ukrainian nationals who have had to flee abroad but use Diia to stay connected to the country. Ukraine's success in enabling the continuity of its digital products during the war showed that the nature of warfare had changed, as according to Maxwell, "We used to assume that this is just how it is in war—everything gets destroyed and you have to rebuild from nothing. But by migrating to the safety and security of the cloud, the government and its citizen services prevail."[28]

The war in Ukraine has taught the world any number of lessons, but the fact that e-government operations could be maintained in part by keeping backups outside of a country's physical territory is sure to be one of the most important. For Estonia, which had long dealt with a dangerous neighbor and repeated occupations, the lesson is one they had also learned thanks to the Russians. The 2007 cyberattack which disrupted Estonian society also served as a wakeup call to the government of both the importance and the danger of running a digital society, but it was Russia's 2014 invasion of Ukraine which showed the urgency of enabling continuity for state operations in the case of a kinetic attack on the nation. This was a serious concern shared by a range of analysts and former government officials including General Sir Richard Shirreff, the former NATO deputy supreme commander for Europe, who went as far as to pen a theoretical scenario of a Russian invasion of the Baltics shortly after his retirement from NATO. Published in 2017, *War With Russia—An Urgent Warning from Senior Military Command* was described by *Politico* as "a military man's rallying cry" and Shirreff says that he hopes the book is preventative rather than prescient and that it "…is not 'fiction as such:' It is 'fact-based prediction, very closely modeled on what I know.'"[29]

Estonia's answer to the dangers it faced, as it so often has been, was an ingenious mix of product and policy design in the form of the world's first data embassy. In June 2017, Estonia and Luxembourg signed a bilateral agreement that would enable the former to host an "embassy" in part of an existing data center in the small nation tucked deep in Western Europe, where physical assault by malign foreign actors is exceedingly unlikely.[30] Unlike traditional embassies known for hosting dinner parties and bloviating bureaucrats, the data embassy would simply be decorated like a regular data center, serving information instead of hors d'oeuvres. However, because of the unique agreement between Estonia and Luxembourg, the area of the data center functions just like an embassy with full immunity.

While the Vienna Convention on Diplomatic Relations provides most internationally agreed-upon rules for diplomacy and embassy operations and may cover such newfangled setups, according to an exploratory analysis co-authored by Laura Kask, former Chief Legal Officer of the Estonian CIO's Office, there is a lack of legal precedent given the concept's novelty which makes the additional bilateral agreement key to the innovative setup.[31] And according to Siim Sikkut, "The agreement that we have with Luxembourg is that it's our territory, our jurisdiction,"[32] which means that the data will remain under the Estonian legal system. This is a critical part of the agreement, one that may have led to some hesitation by Ukraine to allow the movement of government data outside its physical border pre-conflict without similar terms. The data embassy concept seems to have caught on, with the nation of Monaco later forming an e-embassy in Luxembourg,[33] and since Russia's invasion of Ukraine more governments, especially in East Asia, are looking for ways to create backups of government data that aren't physically stored in their territory according to AWS's Liam Maxwell.[34]

With the data embassy launched in 2017,[35] the Estonian government had a solid contingency plan in place that would enable it to continue to provide essential services even in the worst-case scenario where the country was occupied, and the country's democratic leadership once again had to become a government in exile. Nathan Heller's 2017 in-depth *New Yorker* piece on Estonia's digital society detailed how such a scenario could play out:

> 'If Russia comes—not when—and if our systems shut down, we will have copies,' Piret Hirv, a ministerial adviser, told me. In the event of a sudden invasion, Estonia's elected leaders might scatter as necessary. Then, from cars leaving the capital, from hotel rooms, from seat 3A at thirty thousand feet, they will open their laptops, log into Luxembourg, and—with digital signatures to execute orders and a suite of tamper-resistant services linking global citizens to their government—continue running their country, with no interruption, from the cloud.[36]

For some, the idea of running a state from the cloud or a smartphone sounds like science fiction, but for Taavi Kotka and many Estonians, this is a natural evolution of the nation-state. According to Taavi, "Digital countries will be a new normality some time soon," and "The concept of a country has changed. Land is so yesterday. It doesn't matter where you physically live or operate. This is how the game will change."[37] While this may seem like hyperbole and unlikely to come to pass, the increasing interest in charter cities, special economic zones, and more from tech luminaries and crypto investors like Balaji, Thiel, and Vitalik Buterin, the cofounder of the cryptocurrency Ethereum, show it is by no means an impossibility.

This line of thinking is directly aligned with the Country-as-a-Service concept and the tenets espoused in *The Sovereign Individual*. Rather than focus on what the nation-state traditionally has been, look to what technology has enabled states to do (or will soon). In this case, become a state that could operate anywhere in the world, picking up and moving if needed. However, this view somewhat glosses over the deep bond many have to their home, especially those who have fought to defend it. But regardless of where they are physically, Taavi says that Estonians have shown that they can survive in the cloud.[38] Although this is probably more reasonably put as that Estonians' *data* can survive in the cloud. People still require a physical environment to live… at least until Silicon Valley replaces us all with automatons or figures out how to upload our consciousness to the cloud.

Ukraine and Estonia have endured hard-taught lessons which have shaped them into the nations that they are today. For both countries, facing an aggressive neighbor has catalyzed innovation and creativity as they try to secure their own survival. For much of its post-independence existence, Estonia has functioned as an early warning system for the West as to coming Russian tactics for hybrid warfare, prompting domestic investments in cybersecurity, civil-military fusion, and innovative new programs like data embassies. Today, Ukraine bears the brunt of Russia's fiercest attacks, both kinetic and non-kinetic. Like Estonia, the country has shown that it can survive in the cloud (and proven that even during wartime, digitalization at a rapid clip is possible, leaving little excuse for the rest of us). And it's no wonder that Ukraine has been able to weather the storm so well given how diligently the country has absorbed Estonia's lessons and tactics, often working to leapfrog existing Estonian e-government infrastructure and related policy. For the rest of the West, absorbing the key lessons of Estonia's transition to a digital society pioneering new forms of governance thanks to a mentality that holds up citizens as customers and rewards innovation and common-sense reforms will be critical for managing the transition to the information age. This is doubly true given the rapid rise of artificial intelligence which is supercharging the pace of technological progress and societal change.

17

THE EVOLUTION OF THE E-STATE

As part of the adoption (and sometimes a continuation) of a mindset which prioritizes the needs of "customers," public servants in Estonia are frequently examining how to make government more responsive and useful for everyday users. In the early days of the e-state, the answer was digitalizing almost all government services to make them more easily accessible for end users and more streamlined for public servants. Thanks to the e-government infrastructure built over the last several decades, the huge amounts of data collected, and advances in technologies like artificial intelligence, new possibilities for how the government delivers public services have emerged (even as the country grapples with how to maintain now legacy software while trying to upgrade services). For the next stage in the e-state's evolution, an intriguing idea has taken hold: that sometimes the best possible interaction with a government service is one that is proactive, with no action needed from users or public servants. This, combined with leveraging artificial intelligence, and the idea of turning the e-state from product to platform, is one of the ideas guiding the evolution of the e-state.

The idea behind proactive government services (sometimes referred to as invisible government) is that the government will simply anticipate what you need and then do it for you. Ambassador

Viljar Lubi,[1] a longtime booster of innovative ideas and experimentation in the public sector, has compared the idea to how Amazon provides recommendations for users based on purchase history.[2] According to a collaborative white paper titled "Proactive Public Services—The New Standard for Digital Governments" authored by Dr. Keegan McBride, a group of academic and private sector leaders, and the Estonian e-government service developer Nortal, proactive public services "represent the pinnacle of this new wave of digital public services," and,

> can help governments fulfill citizens' increasing expectations that digital public services be provided with the same user friendliness as private sector digital services. It also has the potential to reduce costs and required personnel, freeing up public funds to tackle the most pressing challenges of our time, such as climate change, and mitigating the impact of increasing labor shortages in the public sector, which confront many developed countries.[3]

Proactive public services are part and parcel of the mindset intrinsic to Estonia's development of e-government systems—that the government should act like a business trying to provide value to "customers" and do whatever is necessary to improve the product it offers. It is also well in line with a standard belief among players in the local e-government scene that the number of touchpoints between citizens and government should be minimized, benefiting citizens with higher quality services and a smaller and less costly government. In the e-Governance Academy's 2023 book *Twenty Years of Building Digital Societies*, which shares many of the core lessons learned by digital government developers and policymakers, Siim Sikkut shared his view that,

> Digital bureaucracy is a tool... I don't live my life to deal with bureaucracy, even if it is a comfortable process. The best state is a state that doesn't interfere with me. I get a human touch from other places, with my children and family, with friends at work.[4]

Making services automatic and nearly invisible to users can make their lives easier and save money for the public sector. However, this is an area fraught with potential harms as well as ethical considerations over the delivery of public services and the use of citizen data. As the rollout of proactive public services is still in its early stages in Estonia, many of the dangers are currently hypothetical, but it doesn't take a great imaginative leap to imagine a world in which such services enable a gentle paternalistic authoritarianism. Linking up separate data sources, governments might, for example, start sending users reminders not to buy too many sweets on the next trip to the grocery store if their doctor has written a note about obesity in the electronic health record. However, such issues are generally not related to any specific technology, but instead to the decisions made by elected officials who control the use of such information.

Because of a general wariness towards big and powerful government after decades of Soviet occupation, the Estonian public (and early political leadership) has facilitated strong rule of law and robust oversight of government officials and operations, with citizens easily being able to use simple and accessible legal channels to prevent government overreach. For countries seeking to emulate Estonia's development as a digital society, it is critical to remember that the users of government services need a major say in how they are designed and developed to prevent backlash and create services that are useful to the end customers, not ones that sound good to other government officials. It is also important to remember the context in which such products are built. In some countries, what is acceptable for a government to do proactively for a citizen may be unacceptable in a neighboring nation with a different culture, history, and values.

Despite the potential risks, the allure of proactive public services is strong for both users and the public sector. According to former President Kaljulaid, after independence when the country depended on new tax revenues, it could relatively easily make tax filing a digital service connected to existing private sector online banking services,[5] creating a mostly automated process which helped drive much needed collections. With proactive government services, things are taken a step further—a user needn't even think about taking an action online, it will just be done for them. The potential use cases are near endless, and the Estonian government is just at the beginning of the journey of rolling out proactive services. Existing services include the automatic registration of benefits for new parents and streamlining the process for getting new documents after a marriage if there has been a name change, and many more like a proactive service centered on retirement and care for children with long-term health issues are on the way.[6] One can imagine the difference it makes for a new parent to have government processes streamlined, helping them not only save time but potentially get access to resources they didn't know they were eligible for, a marked difference from many American social support programs that are often built around aggressive means testing and arbitrary hurdles to prevent potential fraud rather than optimizing for conveniently delivering high-quality services.

Given that many leading Estonian politicians still hold the free-market beliefs associated with Prime Minister Mart Laar's government and that of economist Milton Friedman, it may seem somewhat surprising that they would embrace the automation of government services and make it easy for citizens to get access to various social programs. However, a *Wall Street Journal* opinion piece by author and journalist Andy Kessler discussed the use of e-government services and extensively quoted former Prime Minister Kallas, saying,

> Ms. Kallas noted her government uses these digital tools "to decrease, diminish bureaucracy." That's how to create small government. "It's cheaper and our debt is much lower as well." Though it's rising, Estonia still has the lowest ratio of government debt to gross domestic product in the EU.[7]

By effectively collecting and harnessing data in an equitable and privacy-friendly manner, the Estonian government has been able to build a system that minimizes opportunities for fraud while optimizing public services for citizens, making government services more accessible and more efficient at the same time (and setting the government up well to leverage AI, which depends on data to function effectively).

The time (and monetary) savings benefits of proactive public services should not be underrated. By automating processes and eliminating unnecessary work on the part of both public servants and end users, both save time and money or resources. In the U.S., burdensome administrative procedures, like those imposed by the Department of Homeland Security in various ways—going through a TSA security checkpoint at the airport, applying for a visa, and other related activities—led to more than 200 million hours of paperwork borne by the public in 2021.[8] However, after an agency-wide effort to decrease the mountain of bureaucracy, that burden had been lessened by 21 million hours in 2023.[9]

More than wasting citizen's time, bureaucratic processes can lead to those deserving of benefits missing out. According to the 2023 White House report from the Executive Office of the President of the United States, "Tackling the Time Tax," "By one estimate, every year more than $140 billion in government benefits that Congress has authorized goes unclaimed," and,

> One important reason why members of the public do not take advantage of government programs for which they may be eligible are administrative burdens—costs like the "time tax" required to learn about a program, fill out paperwork, assemble required documents, and schedule visits to government offices.[10]

For many public servants and citizens, the status quo is untenable, but there is hope that the use of AI across government can dramatically improve public services, both for back office functions and for creating easier-to-use interfaces like conversational government chatbots.

While much of the world has become wrapped up in the hype from AI-enabled products after the launch of ChatGPT in late 2022, Estonia has long seen the potential of AI for both public and private sector services. For a country that has struggled to compete or stand out globally simply because of its small population size, digital capabilities were first identified by former President Ilves as an opportunity to use as a force multiplier, helping the economy compete globally despite having fewer people making up its workforce. AI is expected to supercharge that ability, allowing small groups to outcompete much larger ones. In the 2000s, multibillion-dollar firms like WhatsApp were built with only a few employees, but now with the advent of AI, it may soon be possible for a single person to leverage AI to create a billion-dollar company on their own. A 2024 report commissioned by Google and created in cooperation with the Estonian government predicted that generative AI could contribute several billion euros or around 8% of Estonia's GDP yearly if the country is able to adopt the technology throughout the private and public sectors.[11] This would be a game-changer for a country of 1.3 million, creating a more dynamic economy and allowing the government to shrink its workforce, creating an even smaller and more efficient government and which may become important even for more populated nations in the West given exceedingly low birthrates and population decline.

According to Marten Kaevats, who started his career as a community organizer before becoming a digital and innovation advisor to the Estonian government under the Kaljulaid administration, the government really started digging into the potential for AI in 2016 as part of a taskforce on self-driving vehicles.[12] Thanks to the work of that taskforce, testing various types of self-driving vehicles has been legal across the country since 2017, spurring the advancement of autonomous solutions like the Cleveron 701, a self-driving delivery vehicle created by a local startup best known for massive towers in retail stores like Walmart used for self-pickup of preordered items,[13] and an autonomous delivery robot built by Starship Robotics which bears a passing resemblance to a Star Wars Episode IV robot seen on the Death Star, albeit in a stark white paint job. In a 2017 blog post discussing the government's approach to AI, Kaevats pointed to the opportunity Estonia had to be a leader in the use of AI, emphasizing the country's potential as a testing ground for new ideas while painting a vision of "Estonia as a pathfinder, constantly moving in uncharted territories."[14]

In a remarkably short time, government leaders like Kaevats and Siim Sikkut coalesced around a concept for how the country would approach the development of AI-enabled public services, naming it Kratt after a local folk tale. The Kratt is a creature formed out of ordinary household materials through a blood pact with the devil and must do anything that its master orders it to; it can only be destroyed by being given an impossible task. Importantly, the Kratt is a tool, and as such can be leveraged for good or for ill and has exceptional power, but if used incorrectly it can cause harm. Nearly all Estonians know of the Kratt, so the government decided to use it as a metaphor for AI and how the government thought about making AI safe for critical use cases. This eventually culminated in the development of the Kratt law which helped to define algorithmic liability (the Kratt law was eventually shelved before passage in light of the larger pan-EU AI Act which covered many of the same legal issues),[15] the ambitious Kratt strategy which outlined how the government would utilize artificial intelligence,[16] and the Bürokratt virtual assistant system for government services.[17]

The government's strategy for the use of AI was extremely ambitious. A 2019 policy document from the Ministry of Economic Affairs and Communications under Siim Sikkut's leadership that outlined the country's national AI strategy for the next two years proposed a host of ideas to speed the update of AI in the public sector including creating an AI-based decision making support system for state institutions and the creation of public sector sandboxes for the testing, development, and evaluation of AI applications.[18] The document also highlighted the government's commitment to invest 10 million euros in the development of such solutions—a not insignificant amount for Estonia's lean government.[19] While advancing public sector services was at the heart of the strategy, Siim also emphasized the potential for attracting new investment and innovation in the private sector by taking a global leadership role in AI.[20]

By 2023, the Estonian government boasted around 100 AI-driven use cases that were already active or in development.[21] According to Siim, AI presented an opportunity to "take our existing digital progress to a whole new level and reap the next stage of efficiency benefits for [the] public sector."[22] Those AI projects have run the gamut from a computer vision project developed to recognize animal and bird species using image and video data from trail cameras[23] to a tool for the Consumer Protection and Technical Regulatory Authority which automatically scraped social media platforms to identify advertisements that didn't meet government standards so a human didn't have to trawl the web themselves.[24]

However, the most ambitious AI project that the country is working on is the Bürokratt system (which is a bit of a pun as the name means bureaucrat, but now AI-related). The program formed a core part of Siim's vision for the future of government services by incorporating proactive public services with all the data the government has collected about a user along with an intuitive and natural user interface all for the purpose of simplifying people's lives.[25] For Ott Velsberg, who has served as the Chief Data Officer of Estonia since 2018, "Bürokratt is not just an IT project but a concept of how digital services and the state could operate in the age of artificial intelligence." After the launch of an initial pilot of the system, Bürokratt is now up and running, although it is still in early days and not operating anywhere near the level of complexity that Siim, Ott, and others hope that it will one day in the future.

The Ministry of Economic Affairs and Communications has created an example of how the system will someday be used in the form of a short dialogue to highlight their vision.[26] The story follows Pille, an Estonian who has planned a trip to Thailand in a month. As the data of her pending trip feeds into the government system, the Bürokratt personal assistant app on her phone provides a notice from the Police and Border Guard Board that her passport expires in six months, but when Pille asks the system to remind her in five more months, the system then alerts her to the fact that the Thai government requires passports to have at least six months' validity when entering the country. Pille then requests that the system order her a new passport, which the platform then complies with after Pille provides her fingerprint and security number along with an appropriate photo. With that, Pille's new passport will be delivered to her home within five business days, potentially averting an expensive disaster at the airport.

For some, the experience may feel invasive or paternalistic, but for many, this sort of proactive service aided by artificial intelligence would be a welcome change from the status quo of interacting with government services today. According to a Deloitte survey on the American perception of government digital services, "Respondents are generally willing to share data for personal and public benefit,"[27] showing that for the right service, many users will happily part with their data. For Velsberg, the goal of Bürokratt is "to offer the best possible digital state experience in order to make communication with the state radically easier for both entrepreneurs and citizens."[28]

As with many other systems, the government has worked to make Bürokratt as transparent as possible, even creating a code repository on GitLab which documents Bürokratt's high-level system architecture[29] and Velsberg follows the north star of "keep it simple, stupid," to make sure that the AI products the government develops don't become complex case studies or shiny objects that fail to deliver for end customers.[30] As time has gone on, the government's ambitions have only grown when it comes to AI. In a 2024 interview, Velsberg stated that "By 2030, we envision a government where every individual has personalized digital assistants for education, mental health and other areas."[31]

While the Estonian government has been a frontrunner in the establishment of AI policies and e-government systems, it has also been buoyed by the uptick in interest in AI due to ChatGPT's meteoric rise. According to Robyn Scott, the founder and CEO of Apolitical, a knowledge-sharing platform for public servants with more than 200,000 global members, a 2023 survey showed that,

> as early as March 2023, 39% of members polled by Apolitical were already using it in their work, but that had jumped to 54% by July, with 22% of their members using it for research and 19% to help with writing.[32]

Apolitical polling has also shown that most public servants worldwide are optimistic rather than pessimistic about the use of AI in government.[33] In the Estonian public sector, there has been a similarly large rise in interest and use of generative AI solutions to augment the existing government AI strategy and current workflows. According to Velsberg, within the next few years the government's goal is that there should be AI implementation in 75% of public sector organizations, but always with a grounding in finding use cases that create tangible value.[34]

For many, the melding of AI and public services in Estonia is a natural fit with the country's role as a place "where stuff happens first."[35] But some, like Professor Joe Burton of Lancaster University, see serious risks in the adoption of AI products by governments, highlighting fears of the militarization of AI and the potential for creating surveillance states.[36] According to Burton,

> The other major concern here is the use of AI by governments in surveillance of their own society... In 2013, it was revealed that the US Government had developed autonomous tools to collect and sift through huge amounts of data on people's internet usage, ostensibly for counter terrorism. It was also reported that the UK government had access to these tools. As AI develops, its use in surveillance by governments is a major concern to privacy campaigners.[37]

Even former Vice President Kamala Harris has talked about the dangers of the technology, saying in 2023 that it "has the potential to dramatically increase threats to safety and security, infringe civil rights and privacy, and erode public trust and faith in democracy."[38]

Despite concerns about the use of AI, many take the view that the genie is already out of the bottle and the only way to combat the dangers of new AI technologies is by using AI. Tech luminaries like investor Marc Andreessen are aligned with this view or even more aggressively in favor of adopting the technology as rapidly as possible, with Andreessen's 2023 *Techno-Optimist Manifesto* emphasizing the need for "accelerationism" in order to invent new technologies which will (hopefully) create a better world for most.[39]

Based on Estonia's rapid clip working with private sector entrepreneurs to develop AI-enabled solutions to tackle some of the most pressing problems coming from AI technology (such as hyper-realistic imagery and video content in the form of deepfakes), it is safe to say that the prevailing view is that new technologies present more of an opportunity than a risk and that governments must embrace a digital-first approach. That view seems to be similarly held by the Ukrainian government, as Vice Prime Minister Fedorov highlighted his vision in a conversation arranged by the Rockefeller Foundation, saying that:

> The war that Ukraine is fighting requires making quick decisions, which is possible with breakthrough technology and the most inno-

vative solutions. Artificial intelligence has helped us during this war, proving once more that a digital state is the most resilient one. Not least because it ensures the operation of public services and the economy. It's the 21st century, so now's the time for AI, UAVs, neural networks, Starlinks, and the swift exchange of valuable information. We are convinced that the future is digital, and it belongs to governments that operate like IT companies—quick, efficient, and agile. Digital governance simplifies the communication between a person and the state, makes services convenient and clear, and destroys any possibility of corruption. Digital democracy will let us include more citizens in the process. We have to build digitally resilient countries.[40]

The initial drive to develop e-government systems in Estonia came from a simple goal: to build the best public services for their citizens given the severe fiscal constraints that they faced. For most public servants and political leaders, the fact that those services ended up being digitalized was simply a byproduct of this being the optimal way to deliver public services—nothing more, nothing less. However, many of the systems that were considered pioneering when first introduced in Estonia in the 1990s and early 2000s are now considered legacy systems or at least legacy interfaces. According to entrepreneur Ott Kaukver,

> I think we have ridden the technology wave now for soon close to twenty years. But I think the next twenty years will be different... In technology years, twenty years is a long time, it's a lifetime almost. We now have legacy systems and new technologies coming on board and we need to change and adapt together with that.[41]

As the Estonian public sector increasingly becomes burdened with legacy software (and sometimes bureaucrats and politicians) many increasingly see the private sector as the best hope for the future. Ambassador Viljar Lubi, then Undersecretary for Economic Development, championed the creation of the Accelerate Estonia program which launched in 2019 under the leadership of Mikk Vainik. With the tantalizing tagline "Let's make illegal stuff legal,"[42] the program was created as an innovation tiger team for the government, building "products" that could be as successful as e-Residency.

Like e-Residency, which had used pre-existing digital government infrastructure to rapidly launch and scale, the hope was that the Accelerate Estonia program could leverage the seemingly more innovative private sector to bring in new ideas that public servants and existing government digital infrastructure could help enable. As of 2024, several projects have gone through the program including ideas for the circular economy and novel uses of biotechnology. However, there has yet to be a breakout success comparable to a program like e-Residency. For many of the concepts vetted, they were only made possible by previous developments in e-government services which functioned as a basic platform upon which new products and services could be bolted.

The idea of leveraging the government as a platform is not new. Under Siim Sikkut's leadership, one of the core agenda items in the government CIO office was to "open up our national digital government stack more so that companies can come and build things on top of it."[43] A major impetus behind opening up the digital government stack is to enable the private sector to build new solutions that the government may be too risk-averse to build, lack the resources to take on, or simply not even consider. The move also highlights an ongoing culture clash between those who believe that the public sector should take a leadership role in innovation versus the private sector. Where Taavi Kotka had talked about e-Residency as an example of government innovation on par with anything imagined by the private sector, former CIO Luukas Ilves has said that "government here functions as a platform. [It] cannot build things on its own. We can manage the data assets, we can orchestrate the services, we have to bring in the private sector."[44] For many, this is a natural progression of the state's role, with a common refrain being that twenty or even ten years ago Estonia was a startup, now it has grown into a scaleup,[45] and with it comes different responsibilities and priorities, namely creating a conducive environment for the flourishing private sector and the foundation and platform for them to do what they do best: innovate.

While in Estonia the debate continues about the appropriate roles of the public and private sector, Ukraine's digital leadership has shown that a country of 40 million can deliver cutting-edge

THE EVOLUTION OF THE E-STATE

e-services even in the midst of an existential conflict. Ukraine's Diia app solves real problems for citizens and contributes meaningfully to daily life, the war effort, and providing a semblance of stability via the "State in the Smartphone" concept. The fact that the Ukrainian Digital Ministry has been able to advance so quickly with so few resources relative to much wealthier nations shows that what truly matters in creating a world-class digital government is mindset. It remains to be seen whether Estonia can manage to maintain the mindset which helped it become a digital powerhouse in a single generation or whether the country will succumb to the weight of legacy systems and bureaucracy.

18

MINDSET OVER MATTER

Thousands visit Estonia every year to try to understand the secrets behind how the country has leapt in just three decades from being a relatively poor, newly re-independent nation to a thriving tech hub with multiple unicorn tech companies and a digital government leader where more than 99% of government services are accessible online. For some, the purpose of the trip is to find out how to create a startup ecosystem; for others, it's a chance to get a peek at what many consider to be the future of government. Many hope to find a silver bullet. Unfortunately, such quick fixes are never forthcoming as the lessons of Estonia are complex, multifaceted, context-dependent, not simple to implement, and most of all require wholesale culture and mindset change across the public sector. According to former Prime Minister Kallas,

> When I speak to people outside Estonia, then I'm often asked what is behind our digital and e-governance success. And I always explain, it's about mindset. We see e-governance not just as a question of building technology, e-governance is also a question of building democracy.[1]

The stakes of being able to create a flourishing digital government and tech ecosystem have never been higher. The global economy is increasingly dominated by tech giants who exert tremendous control over the daily lives of millions. Countries that fail to build their own digital societies and technology capacity and create some measure of digital sovereignty may find themselves dependent on companies with different values, sometimes following different rules, for something as simple as sending a message on a smartphone. The Chinese Communist Party clearly saw the danger of such dependence long ago and has heavy-handedly (and mostly successfully) worked to create a stable of domestic tech champions which they are now using to push their political agenda on the rest of the world in the form of facial recognition technology, social media, and 5G communications infrastructure. Meanwhile, the European Union has mostly failed to create genuine global tech leaders on the scale of countries like the U.S. and China (despite Estonia's best efforts), instead resorting to using regulation in a cat and mouse game with foreign tech behemoths to get them to "play fair," at least within the EU's borders. As the lines between defense technology and private sector technology become increasingly blurred, for example, with AI-enabled drones or advanced energy storage devices, the creation of a thriving domestic tech ecosystem will be more of an imperative than ever for a nation's security.

At the same time, it is clear to most, whether an average citizen consuming government services or a public servant working to provide them, that the way government operates in much of the Western world is fundamentally broken. The systems and processes that were created in the nineteenth and twentieth centuries in leading industrial democracies like Germany, Japan, and America have shown themselves to be painfully insufficient in the digital age. Government bloat, poor user experiences, procedure fetish, an inability to deliver on IT modernization projects—all are symptoms of a legacy bureaucracy that has not adapted to the modern world.

For governments that do manage to adapt, the benefits are tremendous. According to Professor Rainer Kattel and Ines Mergel's 2018 working paper "Estonia's digital transformation: Mission mystique and the hiding hand," "The Estonian government claims that its e-government infrastructure has led to annual savings of about 2% of GDP and more than 800 years in working time for the public and private sectors."[2] If other countries were to implement reforms like those of Estonia, billions of dollars of value would be created. Even better, citizens would no longer be burdened with unnecessary paperwork and processes and the dream of a smaller, more efficient government could be realized.

The human toll of the failing industrial-era system is hard to understate: U.S. veterans unable to prove their eligibility for care, creating needless indignities for those who served; new parents having to deal with piles of paperwork, taking precious time away from their children to jump through arcane government processes; taxpayers not just handing over their taxes to the state every year but having to pay a private sector company for the pleasure of doing so when the government has all the information it needs to pre-fill tax forms for most of the population. Estonia has proven that the way governments deliver public services can change, that the way things are in much of the Western world is not set in stone, and that there is a better way to run government operations—one that is both efficient and democratic.

That human cost also adds up into something much less tangible: loss of faith in the democratic system. Although causality is difficult to demonstrate, there is little doubt that part of the allure of the modern political strongman is that they can promise results after decades of government failures to deliver. There is a reason why Mussolini making the trains run on time has become an enduring meme (despite its inaccuracy). As former U.S. Deputy CTO Jennifer Pahlka has said, "Elites understand policy. The rest of us understand delivery."[3] For a voter who gives a significant percentage of their paycheck to the government, only to encounter inefficiencies, irregularities, and indignities with every public sector interaction, it is no wonder that faith in the democratic system is quickly eroding for many. Wholesale reform and public sector process digitalization represents an opportunity to address the many issues that everyday voters face when they deal with the government.

While interacting with the bureaucracy may never be a pleasurable experience, there is no reason that it shouldn't be as convenient as interacting with a private sector company (ideally one in a competitive market, not a monopoly business like a cable or healthcare company whose customer service may somehow be worse than that of the average government agency). As Marc

Andreessen said in a 2024 conversation with economist Tyler Cowen, "Stagnation is a choice. Decline is a choice," adding that much of the cause of America's recent stagnation and decline is due to the imposition of mountains of unnecessary regulations and restrictions on those who would try to innovate.[4]

Digitalization of government services may be hard, but as Estonia (and others, like Ukraine) demonstrate, it is possible. For some nations, privacy concerns mean that the creation of a mandatory digital identity for each citizen that forms the basis of an e-state is politically hazardous. For others, it's a lack of domestic capacity to develop the digital infrastructure necessary to build a functioning e-government that holds them back. However, time and experience has shown that anyone can digitalize—not by following Estonia's exact playbook and certainly not by copy-pasting Estonian products but by charting a path based on first principles. India, the world's largest democracy, demonstrated that even the largest of nations can create robust e-government systems with its rapid development and rollout of the Aadhaar platform, which Nobel laureate Paul Romer called "the most sophisticated ID program in the world,"[5] and which has enrolled more than a billion residents.[6] And countries as diverse as Ukraine, Benin, and Colombia, along with scores of others have worked with Estonian organizations to successfully develop e-government systems in their countries, showing that Estonia's lessons can be applied in a wide range of contexts when effectively localized.

Each nation will have its own unique pathway to success based on its individual history and culture. Still, Estonia's example offers a range of lessons applicable to all nations about the factors necessary for successful digitalization. For example, factors crucial to success include a bureaucratic and political culture focused on innovation, experimentation, and problem-solving; trust between key stakeholders in the digitalization process; and state capacity to educate and create a society capable of building and using e-services.

In a world of dramatically increased mobility where a few entrepreneurial people can create tremendous economic value from anywhere with an internet connection, the need for bureaucratic reform in the Western world is blindingly obvious. The e-Residency program should be considered the canary in the coal mine for many countries, including much of Western and Southern Europe, given that it has been able to attract thousands upon thousands of entrepreneurs from Mumbai to Munich to Miami to its virtual shores, allowing them to create and manage companies in a convenient, low-bureaucracy environment.

But it is not just e-Residency which has shown that the current trickle of "Sovereign Individuals," whether a freelance software engineer or the CEO of a remote-first startup, could potentially turn into a flood. Dozens of visas catering to these wealthy and mobile citizens have popped up around the world, hoping to lure workers with foreign income sources to their shores. Even cities have started competing for talent. As governments fail to deliver the experiences and quality of services that citizens have come to expect in a world of instant digital and physical gratification delivered via companies like Uber and Amazon, more and more people will vote with their feet and go to a jurisdiction that caters to their needs (although if governments feel significantly threatened by outward migration, it is quite possible that they make it incredibly difficult to exit using economic or other penalties).

While the track record for reform when it comes to digitalization in most Western nations is lackluster at best, by absorbing the lessons of Estonia, countries have an opportunity to dramatically increase the likelihood of success for future digital initiatives. The learnings from Estonia's journey are numerous but there are four takeaways that are important to remember.

First, people are the most important resource a country has. This is doubly true for countries like Estonia, Singapore, or Israel where there are minimal natural resources, and the country must depend on the ingenuity of individual citizens to help build wealth and prosperity. Once leaders internalize that people are far more valuable than any gem or fossil fuel ever could be, the next steps become very clear: educate and then empower the workforce. In Estonia, it was the Tiger Leap campaign which taught a generation the digital skills they would need to become high-skill workers, as well as the scrappy entrepreneurial mentality that would help them become a global bastion for early-stage startups.

REBOOTING A NATION

Of increasing importance for societies in a world of declining birth rates is effectively attracting and integrating new talent. Lee Kuan Yew called the thousands of expat professionals who came to Singapore "extra megabytes in Singapore's computer," adding that "If we do not top up with foreign talent, we will not make it into the top league."[7] Nations that can effectively absorb talent have a significant advantage over those that are closed, something that helped America become a superpower but which is under increasing threat. According to Taavi Kotka in a roundtable discussion with former President Kaljulaid and government digital advisor Marten Kaevats, "Countries are in competition or I would even say constant war [over] who gets the best people and who can boost and grow the wealth of their own people."[8] Undoubtedly, in the age of AI, where single individuals and small teams can create multibillion dollar companies, attracting the best and brightest and providing the education needed to create a high-skilled workforce will be more critical than ever.

Second, bold and decisive leadership is required if a country hopes to make significant strides forward—and leaders must empower and protect those under them to deliver on their lofty goals. From the start of re-independence, Estonia was defined by bold leadership from the highest levels of government. Prime Minister Laar engaged in economic shock therapy, ripping the Band-Aid off in a single motion rather than slowly reforming the economy and working to create an enabling environment for entrepreneurs and innovators. A sense of duty to the nation compelled him to act. He removed subsidies and tariffs, implemented new tax structures, and even sold old Ruble notes in secret to the Chechens in order to keep the country afloat, a ploy which eventually cost him his job as Prime Minister after the controversy led to a vote of no confidence.[9]

Such stories of bold political leadership abound, from making digital identity mandatory to refusing aid when it could have led to the country becoming trapped in legacy IT systems. Similar lessons can be found elsewhere, including in Ukraine, which has shown itself to be a true digital government leader. Georgetown professor Don Moynihan and Gulsanna Mamediieva, former advisor to the Ukrainian Deputy Prime Minister, observed in a 2023 piece titled "Digital Resilience in Ukraine" that Zelenskyy's visible commitment to creating a new empowered ministry focused on digital transformation was key to its success.[10] The President's support of individuals and an agency that others in the government might consider outsiders horning in on their turf or a waste of resources enabled Ukraine to move rapidly ahead when it came to e-government services.

In Estonia, political leaders gave cover to public servants to experiment and innovate, creating an environment that allowed them to build the e-government and legal systems that would propel the nation into the future. With buy-in at the highest levels of government for ideas that many would consider crazy, like the ability to vote online or the e-Residency program, government intrapreneurs like Taavi Kotka and Siim Sikkut could develop products and services that would have been impossible to create in other countries because of varied political concerns. Today, the country saves billions every year thanks to e-government systems, its e-Residency program has brought in more than 200 million euros in tax revenue and helped put the country on the proverbial map, and i-voting has made engaging in the democratic process more accessible for all.

Third, it is critical to get the legal and technical foundation right when developing a digital society. A large part of this comes with the state capacity derived from a well-educated workforce. But as former President Ilves and many others are quick to highlight, a major factor behind the success of Estonia's e-services were a few key decisions, like making digital identity mandatory, which was necessary to move the entire country forward at once and spur the wide-scale adoption of a digital government few would realize they wanted it until they had it. According to Siim Sikkut, who after his role as government CIO has become an advisor to government leaders across the world, technologies like the X-Road and digital identity have been essential for Estonia's development.[11] However, other nations should not copy Estonia's innovations wholesale. Oliver Väärtnõu, CEO of the e-government service provider Cybernetica, is often asked how to build an e-government like Estonia's elsewhere. He says:

I always reply that other countries would not benefit from exactly the same kind of e-government, because each country has its own systems, legislation, culture and ways of conducting business. But indeed, we can export the basic technologies from Estonia and apply them... I still want to emphasize that those technologies do not come in a box—something you just plug and play.[12]

The last and most important factor is the mindset and culture of those building digital government. Marten Kaevats had this to say when discussing lessons for other states building infrastructure:

Throughout the process, keep in mind that digital is not the goal; a change in mindset and culture is the aim. Many governments and large-scale organizations misunderstand digital as a set of trendy gadgets necessary to get either votes or revenue. Navigating the uncharted waters of digitization means going through rough seas and reefs on an everyday basis. When you discover yourself, yet again shoveling water out of your dinghy, try to remember why you set sail in the first place. Think of digital transformation instead as a tool to help us achieve societal change at scale[.][13]

Scores of other government and private-sector leaders have expressed similar sentiments again and again, from Prime Minister Kallas highlighting the importance of a digital-first mindset and culture to entrepreneur Sten Tamkivi discussing "Lessons from the World's Most Tech-Savvy Government," in an op-ed where he said: "Still, Estonia is a start-up country—not just by life stage, but by mindset. And this is what [the] United States, along with many other countries struggling to get the Internet, could learn from Estonia: the mindset."[14]

And while the concept of culture and mindset is inherently murky and context-dependent, the Estonian cultural north star has always been to solve real problems for real people and use common sense. When former President Ilves was asked how he got the whole country to go along with so many major policy changes, he responded simply that, "the key, again, in retrospect, is that you provide services that people like."[15]

Estonia's experience is not the only example of the importance of mindset. Aneesh Chopra, the first Chief Technology Officer of the United States, talked about the importance of adopting a new mindset in the book *Innovative State*, saying,

> the government has an opportunity to return to its role, from earlier stages of American history, of introducing new services and techniques to the private sector. It all starts with the government embracing a need for a new approach, a mindset that is consistent with the American legacy,[16] and, "We needed to progress from reactive to proactive government. For that, we needed an innovation mindset and strategy, every bit as much as private sector companies did."[17]

Mike Bracken, the UK's former Chief Digital Officer, highlighted the importance of mindset in a 2024 *Financial Times* op-ed in the wake of the Post Office Horizon scandal commenting, "The Post Office-Horizon affair is a human tragedy and a humiliation for the British establishment. Without a radical change in mindset at the top of government and the civil service, it will happen again," later highlighting the need to "redesign government so that it can redesign government services."[18]

It is important to remember that any nation can take Estonia's lessons and undergo the requisite shift needed to create a digital government fit for the era of artificial intelligence, drone warfare, and geopolitical uncertainty. While change always brings risk, what is even more dangerous is maintaining a status quo which has already shown itself to be a clear path to failure for the Western world as voters continually search for states with better capacity and services or the leaders who promise to deliver them, no matter the cost to democratic values. The world is undergoing a tectonic shift thanks to the proliferation of the internet, smartphones, and emerging technologies, and those who don't move forward risk being buried in the rubble, relegated to the dustbin of history.

Estonia's story shows that in a single generation a country can go from being poor with few prospects and natural advantages to being a global leader. All it takes is bold leadership, investing in people, creating a welcoming legal environment, and investing in the right infrastructure for the foundations of the new digital state, while embracing a mindset shift that prioritizes innovation and progress over stagnation and maintaining the status quo (a little luck helps too).

REBOOTING A NATION

For those who find themselves in difficult situations, they would do well to remember possibly the most important lesson of Mr. Estonia, President Lennart Meri. The situation may be shit, but this is our fertilizer for the future.

PART III

LESSONS LEARNED

Mindset over Matter: If there is one lesson to take away from this book, it is that anything can be achieved with the right mindset. While shifting one's mindset may seem like advice given in a self-help seminar, the evidence is overwhelming that to accomplish any long-lasting change, a mindset shift is required. Such a change can start small with a single individual in a single department working as an intrapreneur and infecting others, but ideally the mindset shift is embraced from top to bottom, with executives just as bought-in as everyday public servants and end users. For leaders, it is important to provide a vision for the future that can serve as both inspiration and a north star for those in bureaucracy to work towards and to make clear that innovation, even when it ends in failure, will be rewarded rather than punished. People are the greatest asset of a nation: create an enabling environment that empowers your people.

There Are Many Pathways to Digital Enlightenment: Estonia's journey has been a unique one. It is a small country once tucked behind the Iron Curtain that is impossible to place into a simple taxonomy of Baltic, Eastern-European, post-Soviet or even Nordic: all these labels fail to encapsulate the unique culture of a nation that has defied every expectation to become a global leader in the information age. Estonia's journey will, most likely, not be the one taken by other countries on their road to creating a digital government and society—in the most basic sense because technology has changed dramatically since the country gained independence and newcomers should focus on the next generation of technologies and interfaces and look to leapfrog Estonia, rather than play catchup. However, basic principles like investing in education to create a high-skilled workforce, empowering entrepreneurial public servants, engaging in bold leadership and reform, and promoting a progress-oriented mindset are universal.

Cruise Missiles Can't Take Out the Cloud: Technology and warfare are fundamentally changing the nature of how government operates. In 2007 Estonia experienced the beginning of a new era of warfare, this time in cyberspace. In 2022 the second invasion of Ukraine showed that government could operate in the cloud, even when it was physically under siege. That a country can now operate from anywhere in the world as long as they have access to enough data storage will change the future of the nation-state. Even if Russia conquers all the territory of Ukraine, the country will live on in exile, able to ensure continuity of a range of services by digital means until it can reconstitute itself physically once again.

Don't Buy When You Can Build: State capacity, not just in being able to procure the right solutions from the private sector, but being able to build the products and services necessary for the functioning of the state when needed is critical. For Estonia, this was most clear in the development of e-government services, but it holds equally true for defense and critical industries. Dependence is dangerous, especially when it leads to a nation losing the internal knowledge to reconstruct systems or industries that have been outsourced, as evidenced by America's difficulties with rebuilding basic manufacturing capacity after decades of offshoring.

RECOMMENDED READING

Estonia: A Modern History by Neil Taylor.
The Small States Club: How Small Smart Powers Can Save the World by Armen Sarkissian.
Twenty Years of Building Digital Societies by Peeter Vihma for the e-Governance Academy.
Estonia's Digital Transformation: Mission Mystique and the Hiding Hand by Rainer Kattel and Ines Mergel.
Digital Government Excellence: Lessons from Effective Digital Leaders by Siim Sikkut.
Recoding America: Why Government is Failing in the Digital Age and How We Can Do Better by Jennifer Pahlka.
The Sovereign Individual: Mastering the Transition to the Information Age by James Dale Davidson and Lord William Rees-Mogg.
The Entrepreneurial State: Debunking Public vs. Private Sector Myths by Mariana Mazzucato.
How to Make an Entrepreneurial State: Why Innovation Needs Bureaucracy by Rainer Kattel, Wolfgang Drechsler, and Erkki Karo.
Snow Crash by Neal Stephenson.
"Estonia, The Digital Republic" by Nathan Heller for *The New Yorker*.
The Road to Freedom: Estonia's Rise from Soviet Vassal State to One of the Freest Nations... by Matthew D. Mitchell, Peter J. Boettke, and Konstantin Zhukov.
From Third World to First: Singapore and the Asian Economic Boom by Lee Kuan Yew.
The Baltic Revolution: Estonia, Latvia, Lithuania and the Path to Independence by Anatol Lieven.

RECOMMENDED READING

The Forest Brotherhood: Baltic Resistance Against the Nazis and Soviets by Dan Kaszeta.
Infomocracy (Centenal Cycle) by Dr. Malka Older.
"Culture Eats Policy" by Jennifer Pahlka for the Niskanen Center.
The Big Con: How the Consulting Industry Weakens Our Businesses, Infantilizes our Governments, and Warps our Economies by Mariana Mazzucato and Rosie Collington.
The State in the Third Millennium by Prince Hans-Adam II.

SELECTED INTERVIEWS

STEN TAMKIVI, INVESTOR AND ENTREPRENEUR, ON SKYPE, ESTONIA'S UNIQUE STARTUP CULTURE, AND TECH ECOSYSTEM DEVELOPMENT[1]

You've had an incredibly successful career in tech as an early employee at Skype, EIR at Andreessen Horowitz, and as a founder and investor, but you've also been involved in government from a very young age. Can you talk about how that happened?

I was still in high school when Toomas Hendrik Ilves and Jaak Aaviksoo came up with the Tiger Leap. They formed a committee for the program and at some point realized that they had all kinds of smart people in the room, but no students. So, then they were scrambling to find a student and because Jaak Aaviksoo was my father's classmate in university and they lived next door he basically called me up and said, "Hey, do you want to contribute a student voice to this?" That's how things in Estonia usually happen, you live next door or something, and in retrospect it was quite funny because there were all these people with names that you read in the papers and I showed up to these committee meetings but I had very little to add as a 17-year-old. I was basically listening and observing...

The only formal job I've ever had with the government was in parallel to Skype. Then President Ilves was touring tech companies throughout his tenure and during one of those visits he told me he'd like to have an advisor on digital technology and entrepreneurship topics and I agreed to think about who it could be. I ended up chatting with his chief of staff about these candidates and somehow it came to why don't you do it? I hadn't really considered it but we worked out a way where I had a contract, the only contract I've had with government, as a 0.1 workload position [to nominally be on the payroll as an advisor, analogous to the "dollar-a-year" executives in the U.S. who supported the government during WWII] which was basically a formality.

Can you speak about the impact of Skype on Estonia culturally and how it's affected entrepreneurship in the country?

You can look at the Baltic states as a test group of similarly sized countries with the same or similar history, level of education, all of that, but somehow a lot of things happened slower or still haven't happened there [in the other two Baltic countries]. I think that one thing is that people in Estonia didn't move away... Some other countries getting out of Soviet occupation, the technologists would have moved to Western Europe or the U.S. or whatnot, like Ukrainian developers in the U.S. or Lithuanians in Ireland and in the U.K. In Estonia, they found something interesting to do locally. Proximity to Finland maybe meant that they had some foreign customers or whatever it was, but fewer people had to leave in the field. Then once you have the Skype experience, that means okay, now it's proven that you can build a global thing without being anywhere else...

I have a friend from early Skype, Ott Kaukver, who has said that "Skype is the most successful business school in Estonia." So, you have a few thousand people who have had this direct experience and you can stay here, build something that affects hundreds of millions of people around the world. You learn hands-on about scaling. You see all the craziness, companies being bought and sold, CEOs coming and going, and get the exposure to see this is what European investors are like, this is what U.S. investors are like... Secondly, I think you have people here who made money or they know people personally who made money. It was very hard to come out of Skype and then go work for a ministry, and if you either start a company or join an early-stage company, you can sort of call up your network and say, "Okay, I need half a million to get going." I think the third thing is that we were lucky that Skype was a consumer brand, because you can very well imagine that we would have built an enterprise company... What happens with the consumer brand is that internally, everyone has direct exposure to the product and they know what you're talking about. The parents of kids who say, "Hey, I'm going to work for a startup," they know that, okay, Skype was a startup, it was successful. Internationally, you can use your CV with a line about Skype and open any door in the world and say that you're building the next thing and you have the credentials because people have this direct exposure to the Skype brand. It's interesting, if you look at Lithuania which now has Vinted and Nord VPN (two first unicorn tech companies). If you compare the visibility or direct exposure of those companies to the general population, Skype was clearly different.

My impression from the Estonian tech mafia and the Skype mafia is that entrepreneurs in the country do an exceptional job supporting one another. Can you talk a bit about the local culture in the tech community compared to what you saw working in Silicon Valley?

Well, I do believe that Estonian founders are irrationally patriotic. It's mostly because my generation still remembers the Soviet occupation and so having your own state, it's a special thing. I think that's one huge difference between Estonia and the U.S., or being in those two places, which is like, the country really feels like it's yours. Like if you look at this societal friction and objection that the average American has about the federal government having a unique ID for them. Versus in Estonia, the government is not the others, the government is you. That changes the perception a lot and I think that translates that also if you live in London or in San Francisco, you can be building a very global tech company, but if there is a Slack channel somewhere where you can help other Estonian founders in Estonian, which is our little secret language, there is something special and people do that.

INTERVIEW WITH STEN TAMKIVI

Are there any parallels between the building up of Estonia post-independence given the dramatic resource constraints the country faced and building up a startup, which are by nature resource-constrained?

There is a lot of pragmatism which comes from lack of resources of several kinds, which coincidentally is what startups are. You can find the same in the Icelandic startup scene. I think partially it was the reset at the end of occupation, the poverty and all that, part of it is just like a small country. My overall view on e-governance is that smaller countries have a massive advantage and that's why you see a lot of governance innovation in digital societies happening first in places like Estonia or Singapore or Luxembourg or the Netherlands or wherever. The power distance is lower. You have way more horizontal collaboration between people where if an entrepreneur has an idea they can call up the Prime Minister and make things happen. I don't mean corruption, because it's your classmate. You are making the country better for everyone, together, not asking favors for yourself.

If you're a country that wants to emulate Estonia's success in building a startup ecosystem, are there any recommendations you have of things to do? How about things not to do?

What not to do is easier. Don't go out with a press release that you "will be the Silicon Valley of X." This is the most idiotic thing and nobody believes those press releases, anybody who says this you know they're not real. I think the misconception there is that people try to replicate, but every startup knows that the only way you can compete with an incumbent is that you have to be 10 times better. So you have to find something that you are doing differently and change the game, rather than replicate and be 10 percent better than the Silicon Valley that is already there. The other thing is I think governments usually think we're going to go and create a tech scene. No government can create an economy. You can lay the boundaries and create environments, but people who build companies will decide what they will build and what they will do... So, some governments do stuff like they start giving startups money and a government bureaucrat picking which company will win and which one will lose is an awful idea.

So, how I would approach it is to look at the broad things which are too big for any particular company to solve or any NGO to solve. Good education, real sciences, availability of tech in schools, basic safety nets in healthcare, safety in streets, police and firemen. That's the level where it makes sense to pull together state budget resources and so forth. For everything else that is more specific regarding building companies, the mindset shouldn't be how does the government actively create things but instead, how can the government remove obstacles. So if you look at immigration, how can we remove bureaucracy? How can we create the sort of systems where it's easy for talented people to move in and out? If you look at taxation, how can you reduce burdens for early stage companies rather than throw tax money at different players? If it's basic business bureaucracy, there's pretty adequate data about how many minutes a day, a month, or a year, an entrepreneur spends interfacing with the government in different countries. And still there are countries that think it's fine that it's hours and hours a day. So deal with that rather than creating a startup Silicon Valley...

The other thing is that it's kind of like the grind of building a company. Most of the improvements that you make are miniscule. They're small, iterative improvements towards building a better system. Very rarely do you have a startup and an idea and then you make one big bet and have one PR splash and then you have a successful company. I think extrapolating this to building a startup scene is kind of the same thing. So it's more about how the government builds a feedback loop where they fix a few things. I've had that argument with the Estonian government as well... I think governments often operate on a pipeline of things and they check something off and now it's done. You need to keep learning and iterating. As you know with the e-Residency program, if you gather a bunch of people who have moved to Estonia every six months for a meeting, you get a hundred new small things that could be fixed. So how do you create the mechanism to plow through that?

EXPLORING THE UNIQUE LINKS BETWEEN ESTONIA AND FINLAND

DR. MATTHEW MITCHELL, SENIOR FELLOW AT THE FRASER INSTITUTE, AND CO-AUTHOR OF *THE ROAD TO FREEDOM: ESTONIA'S RISE FROM SOVIET VASSAL STATE TO ONE OF THE FREEST NATIONS ON EARTH*[1]

Finland would play a crucial and immediate role in Estonia during the 1990s, and the lack of such a benevolent uncle for Latvia and Lithuania probably explains why Estonia was able to Westernise more quickly.

Neil Taylor, author of *Estonia: A Modern History*[2]

What was the impact of the Communist occupation on living standards in Estonia compared to neighboring Finland, which had managed to fend off the Soviet Union?

I think it's pretty obvious that communism didn't deliver on its promise of material abundance. You can see this in any number of ways, but the starkest is just to look at what the average Estonian was earning compared to the average Finn before and after the socialist experiment. In the decade before the socialists took over in Estonia (just before WWII), the average blue-collar Estonian worker earned 84 percent more than the average blue-collar Finn. Then you fast forward to 1988 and you've got a half century of socialism. Socialism was supposed to out-produce capitalism. But now, according to the official exchange rate, the Estonians earned just 22 percent of what the Finns were earning.

But income isn't everything. We also care about other things like, how much can people afford after they work? The average Estonian and the average Finn's purchasing power were comparable before socialism in Estonia. The average Finn could afford a little bit more sugar on his salary compared to the average Estonian, who in turn could afford more white bread, but after socialism it was just extremely lopsided. The average Finn could buy seventeen times as much coffee, and three times as much butter and pork and milk and cheese and beef and sugar. They had to work significantly less in order to afford things in Finland relative to Estonia.

Can you talk about the deep cultural similarities between Estonia and Finland that has led Estonia to long use Finland as a form of "measuring stick" for economic and social progress?

One thing that's interesting is they share the same Finno-Ugric language group. There's not a lot of people on the planet that are a part of this language group. They both have a long experience with political domination from the outside. It was not all bad. For example, the experience of Swedish control in both countries from what I understand was actually generally pretty positive. Both countries still celebrate the legacy of property rights and economic freedom that the Swedes gave them. They both had experience with Russian rule, they both gained their independence after the October revolution. They both had parliamentary democracies, they both had joined the League of Nations, and had strong Western-oriented ties. Similar geographies, disease environment, climates, all of that makes for an interesting comparison.

There are sometimes jokes about the link between Estonia and Finland being forged in no small part by vodka tourism from the latter to the former but clearly the cultural ties run deeper— are there any examples that you can highlight to show the deep links between the nations?

I think one way to look at this is during the brief period when the Nazis had taken control of Estonia during WWII and their efforts

INTERVIEW WITH MATTHEW MITCHELL

to try to recruit Estonians to the cause. From our book, we talk about how few Estonians volunteered. Then the Germans turned to compulsory conscription, but they still only netted about 5,300 men for the Estonian Legion. An almost equal number crossed the Gulf of Finland, which obviously is pretty dangerous during wartime, so that they could fight the Soviets under the Finnish rather than the German command. So, I think there you have a good example of this long history of Estonians being willing to go and help the Finns and fight the Soviets under their flag rather than under the flag of Germany. So, I think that those cultural and political ties definitely run deeper than just liquor. And you've got that long shared history of experiences which helps bind people together,

Can you speak to the friendly competition between the two nations and how that may have impacted Estonia's development?

Across the world, for example in the United States, South Carolina is emboldened by changes made in Florida. And this is just a normal fact of life. Part of it is that another jurisdiction doing something first provides a proof of concept. The term "leader" connotes the idea that a policymaker is at the forefront. But I think political economy suggests most of the time that (political) leaders are followers. There's the rare one that is a leader but most of the time they're extremely risk-averse based on their incentives. But you can battle that risk aversion if there is an example. Obviously, we think of the first mover as having this great advantage because they're first on the scene, but typically the second mover is in a better position because they haven't risked as much capital, and they can learn from the mistakes of the first mover.

With the Finnish example, and I think you describe it correctly, with this idea of it being part friendship but also part rivalry as well. They can do it and we can do it too. In technical matters, the Finns rolled up their sleeves and helped the Estonians right after independence in establishing banks with modern technology, so they went straight from a cash economy to an electronic economy and skipped checks. A lot of the shops that were established in Tallinn right after

INTERVIEW WITH MATTHEW MITCHELL

independence were either owned or at least had capital from Finland. So, I think the rivalry and the friendship both helped sustain Estonia during and after the long socialist experiment.

I'd like to end the interview with an excerpt I found particularly telling from The Road to Freedom:

> In the end, the 50 miles of water between Tallinn and Helsinki was not wide enough to keep all goods and ideas at bay. The geographic gulf was narrowed by shared historical, cultural, and linguistic bonds between the Estonians and the Finns. Though the Finns were always careful not to antagonize the Soviets, they did their best to keep these bonds strong.

GETTING THINGS DONE IN GOVERNMENT WITH MARINA NITZE

A DISCUSSION WITH THE CO-AUTHOR OF *HACK YOUR BUREAUCRACY* AND FORMER CTO OF THE U.S. DEPARTMENT OF VETERANS AFFAIRS[1]

The first time you're doing something in a bureaucracy is always going to be the hardest and longest, often by 10 or 100x.

Marina Nitze

What is bureaucracy hacking?

Bureaucracy hacking is about using the rhythms of the bureaucracy and the process of the bureaucracy to make it work better for your intended outcome. Everything is a bureaucracy. But you can be an effective bureaucracy. Bureaucracy does not mean that you have to be ineffective.

Let's zoom in on a real-world example of bureaucracy hacking. By the time you left as CTO of the VA, more than 600,000 veterans had enrolled in healthcare from their mobile phones and over 1,500 disparate websites were consolidated. How did you make such dramatic progress?

It's a complicated answer, but I would say that it started with a vision. It's not about having a thousand websites. It's about having a really high-functioning experience for veterans. So, it really started out with listening to veterans, sitting on park benches at the hospital, sitting in waiting rooms, hearing their experiences. And consistently, they were terrible experiences, and not only just terrible, but very disjointed. So, we had a thousand websites, but it wasn't like we had a thousand good websites. We had a thousand websites that were written in legalese, that were confusing, that had conflicting information. I started trying to fix it the wrong way, which is often when we want to tackle these huge problems, we start with the biggest. Instead, the way that I was able to ultimately build a central digital experience was building Vets.gov on the side to start. I didn't try to take over VA.gov, so really, I made a thousand and one websites and from that beginning, looked to see who around here is willing to join my website to consolidate. There were not very many people in the beginning. It was the Office of Economic Opportunity, so we put the GI Bill comparison tool calculator up and then we worked with the National Cemetery Administration, then the board of Veterans' Appeals and we slowly built that political capital and trust with people.

Lesson: Start small, work to build trust and credibility, and add value fast.

How did you bring people along with your vision?

In the beginning, anybody that was willing to talk to me, I was willing to talk to them and I was willing to help them solve their problems, even if I didn't see any clear path at all to how it was going to help me. It was really about just being a humble human and saying, how can I help? I can make Excel spreadsheets; I can work on macros. These were things that were (heavy air quotes) below my pay grade. I just don't believe in that concept. Anything I can do to be helpful; I'll be helpful. I want to build genuine relationships with you so that you get to know that I am a decent person.

Shouldering risk was a really big part of it. I'll sign paperwork, I will personally take the responsibility. If there's an investigation, I will go and I will say, I did this, and this was my decision. Especially early on, that's really important. And you want to shower people who do the thing you wanted them to do with credit. In my case at the VA, I would take the time to write letters to their managers and get them promotions, I would email Secretary Bob (Robert McDonald) and CC their manager... Anything that you could do to say, I want your peers to see that this is behavior that is rewarded, such that you keep doing it.

In Hack Your Bureaucracy, *there's a section titled "Don't Aim for Culture Change." Can you talk about how you see the role of culture in government or other organizations?*

I think a government agency's culture and any organization's culture are a reflection of their risk and incentive frameworks. People's behavior reflects what they believe they will be rewarded for, what they believe that they will be punished for, and the actions that they take fall within that framework. So, you can have a billion hurrahs and motivational speakers and posters about culture change but if people believe that doing a slightly different or new thing will get them in trouble or shamed or not respected by their peers, they're not going to do any of those things.

So, what instead you have to do is demonstrate the change you want to see. That, to me, is the only way you get culture change. What's the change you want? Why aren't people doing it today? And it's never because they suck and are evil. That would be an easy problem to solve. The reason is always a lot more complicated, it's about something in their position description or their performance review and the way that promotions are given out or the way their manager reviews progress. You have to fundamentally change those things to change that behavior.

I don't think anybody's behavior really changes in response to a motivational speaker, no offense to them, but it's not how you make culture change.

HIGHLIGHTS FROM A DISCUSSION WITH FORMER U.S. DEPUTY CTO JENNIFER PAHLKA AND AUTHOR OF *RECODING AMERICA* ON STATE CAPACITY[1]

Changing culture is a very, very hard thing to do... you cannot legislate culture change.

Jennifer Pahlka

What is state capacity?

So, state capacity is, very simply, the ability of a government to achieve its policy goals. You could also say it's the ability of a government to do what it says it's going to do.

What are the common arguments or rebuttals against increasing state capacity?

Very few people are intentionally refuting it on those terms, like let's not have state capacity. But there are certainly concerns about building a government with the capacity to do things whenever the other party might be in control, particularly folks who are afraid of what Trump might do. In the '90s and 2000s, when you had a kind of liberal vs. conservative and regulatory vs. deregulatory divide, you saw the right not wanting the government to have the ability to reach into people's lives and businesses and in a way that they thought was inappropriate. Intentionally obstructing state capacity is one way the right has tried to limit that reach. I argue that crippling the government from being able to act efficiently and effectively actually increases the government's level of intrusion, so I'm not sure they got that right.

You hear very often (and I mean this with heavy air quotes here), "government will never be good at digital. We shouldn't try. We should just try to hire better contractors." I don't really buy that argument either, that we shouldn't have state capacity in digital because we can't be good at it. It's a self-fulfilling prophecy.

Why is state capacity important and how can we strengthen it?

The crisis of state capacity underpins all other crises that we're facing as a nation that are much more visible… There's basically three ways to increase state capacity. You can have the right people, you can focus them on more of the right things, and you can burden them less.

What does it take to meaningfully increase digital state capacity across the public sector?

It starts with recognizing this as the crisis it is… You have to have someone who's very high up, far higher up than we have today, naming this as a top priority in order to set the conditions that we need for change. In peacetime, you get sort of a drift towards proceduralism. In wartime, you cut that shit out and you get shit done. Our [current] message is that it's peacetime. But threats like climate change aren't going to wait while we politely navigate mountains of procedure.

You have a quote in your book, "Elites understand policy, the rest of us understand delivery." Can you elaborate on why it's important for elites to also understand delivery?

I think FAFSA, what's going on at the Department of Ed right now [referring to a 2024 crisis in which tens of thousands of low-income kids were left waiting for their forms to be processed to receive financial assistance], is a very clear example of this. The White House has been crystal clear that they care about low-income kids. The administration has spent so much political capital on trying to get loan forgiveness, and yet, what low-income kids applying for college are feeling right now is utter, brutal betrayal because they are experiencing the delivery instead of the policy. We spent the majority of leadership cycles in that department on policy without recognizing that to help these students, we needed to get the delivery right.

In cases like this, who should be blamed? Where does the buck stop?

The system is bad. Blaming actors for not wrenching victory from the jaws of defeat when the system is going to defeat them is sort of pointless... but on the other hand, somebody does need to be responsible... I don't think a strong leader can fix it overnight, but I do think what our leadership needs to say is: This is a crisis, let's act like it. What do we need to do to hire the right people, focus on the right things, and burden them less? Let's do those things—and fast.

Why does culture matter in government and what is the interplay with state capacity?

A good example of the principles of "culture eats policy" is that all of our laws are designed to create a meritorious hiring system, but in practice, they create a very unmeritorious and slow and ineffective system. So, the policy says one thing, but the outcome is another entirely. If you have an extremely risk averse culture that drives towards maximalist interpretations of anything, then you're pretty much guaranteed to flip the intent by the time it hits the field level.

That gets to the role of oversight, which we really use to find things that went wrong. That alone explains a lot of the risk aversion in the bureaucracy. If you want a bureaucracy focused on outcomes, you need to spend at least 50%, maybe more, of your oversight function showing what good looks like, what the desired behavior is, and rewarding it.

INTERVIEW WITH JENNIFER PAHLKA

Do certain government mindsets play a role in the failure to build state capacity?

The core mindset is one of delivery over policy. The most fundamental thing is moving from this… [current waterfall approach of industrial-era government] to internet-era thinking with a circular, iterative, feedback loop instead of a cascade with one-way, hierarchical thinking.

How can we counter the "accountability double-bind" which hurts public servants when they try to be innovative or do "the right thing" but stray from the exact letter of the law?

Put flaws in context. Is it a very minor procedural flaw that doesn't make any difference? Then don't write it up! Hold people accountable to outcomes instead of process. I think in general the legislature tries to hold people accountable to outcomes without understanding, and doesn't seek to understand, the ways in process forced that outcome. What the legislature needs to do is to say, "what can we do to make it easier for you to get the outcome desired? What is standing in your way?"

HOW THE NEXT GENERATION OF POLICYMAKERS VIEW THE FUTURE

A DISCUSSION WITH MP HANAH LAHE, ESTONIA'S YOUNGEST MEMBER OF PARLIAMENT[1]

My generation's heart aches for Ukraine, not because we lived through a war, but because we have the negative imprint of Russian invasion from our heritage. It matters to everyone who is Estonian how things are going in Ukraine. There really isn't another option.

Hanah Lahe, interview with *The Guardian*, 2024[2]

Can you provide a little background on yourself—I'm especially curious about what motivated you to run for office and what got you into politics originally?

My story actually starts in 2019 when I was younger, I'm 24 now. I didn't care about politics a lot, it wasn't really talked about in my family a lot, maybe just a few curse words here and there... So, I grew up knowing that politics was something dirty that important men in suits do. In Estonia, you can vote at 16 and run in local elections at 18. This had entered into force in 2017 when I was in middle school, so I already had the possibility to go vote, but I didn't because there was no one in my area that I could relate to. No women, no younger people, no one talked about environmental issues, climate, human rights, youth issues, and the same goes for European Parliament elections in 2019 and in the same year we had national parliament elections as well. So even though I had the chance, I never went to vote.

But in 2019, I was working at my first job, which was in a bank, so I was in an economics-related bubble and I needed to be much more in touch with politics as well. It was a good time for me to start learning about politics because the Reform Party won the elections in 2019 but a very conservative and I would say populist coalition was formed behind our backs, leaving us in the opposition. Even if at that time someone wasn't involved or didn't care about politics and wasn't looking at the news, they would have known what was going on because of everything that was happening. Insulting our allies, our neighbors, wanting to leave NATO, EU, exit the Paris Agreement. It was just horrible, and I felt like I had to do something...

My main reason [for getting involved in politics] was no representation for young people, no one talking about climate issues, and really wanting to fight populism.

Can you tell me more about the specific issue set that you care about? We talked a little about climate change but what are the other kinds of youth issues that you think are being neglected?

Well, when it comes to the youth representation in politics, I think it's a really crucial thing to talk about because when we look at the statistics, for example, here in the Estonian Parliament, we have 101 members of which only two are under the age of thirty, and I'm one of them. In the European Parliament, 705 members, with six people under the age of thirty.

They're becoming more aware, especially in these crises with climate, human rights, and I think young people are just scared for their future and they really want to have a say. But when we look at who the decision makers are, then young people are not represented... So it's a question of representation and from that comes dealing with youth issues. I would say the biggest things are climate change, human rights, and AI/digitalization, education, and informal education as well when it comes to youth councils and different study programs... Then the job market, real estate market,

INTERVIEW WITH HANAH LAHE

you know, how easy or hard it is for a young person to start their independent life. I would say it's quite hard.

I'm a little surprised at the lack of youth representation as it seems like during the Laar administration for instance, a lot of focus was on the fact that he was in his thirties and many ministers were younger, and that this youthful exuberance created tangible benefits for Estonia.

Yes, but the same people who were elected when they were really young, many of them are still in the parliament. So, we need this, in Estonia we say "blood change," which I think is happening. We need to fix or improve the voting turnout as well but there are just two people under the age of 30 and I think in the last term it was the same.

I'd love to hear your vision for the future of Estonia or what you hope the country looks like in 2035 or 2050.

For 2035 or 2050, I hope that the coalition we have right now will stay in place for the next three years until the next elections. I hope that in the next elections EKRE [far-right party] won't get to be in the coalition and I really hope that the far-right populism will fade, not just in Estonia, but in general in the world… Here in Estonia and I think the Baltics generally, and maybe I would say the Eastern European countries really look up to Scandinavia and the northern countries in terms of their climate policies and their targets and how they have reformed all of their sectors. I think Finland wants to go climate-neutral by 2035 and Sweden by 2040, so they are a lot more ambitious than the EU generally is and I think this is something Estonia could do as well. Obviously right now we have set the target to 2050, but when we look at what scientists say about the climate crisis, you look at the weather outside now, we really don't have much time left.

A DISCUSSION WITH SIIM SIKKUT, FOUNDER OF DIGITAL NATION AND FORMER CIO OF ESTONIA ON THE IMPACT AND POTENTIAL OF AI IN GOVERNMENT[1]

> *E-governance is not just about building technology but also building up democracy. When the government is not online it just alienates from the people. We need to deliver what is best for the citizen. The future is about personalized and seamless services, leveraging artificial intelligence.*
>
> Prime Minister of Estonia, Kaja Kallas, opening remarks at the 2023 e-Governance Conference[2]

One of the key initiatives during your tenure as CIO of Estonia was an AI system called Bürokratt, which functioned as a new interface for users to deal with the government. Can you talk a bit about what Bürokratt is and the problem it is trying to solve?

When I stepped into the government CIO role in 2017, AI was basically in our pockets. This massive outburst was finally happening in that field and so I quickly realized we needed to catch up. That is why we immediately made AI uptake one of our priorities for my term... Actually, Bürokratt itself is not my idea at all. It was something that Marten Kaevats [Digital Policy Advisor for the Government Office] came to and said let's have an initiative where literally AI and machines take care of the bureaucrat's job and become the interface [between government and users of government services]. The idea made sense but didn't get going for some time. Until we took the lead of it with Kristo and Ott who were the CTO and CDO of the government, and we literally said, "okay, it's a nice idea, but let's now operationalize that…" So, I guess our contribution was that we took the idea and made it into an actual initiative. And that's the vision paper we pushed out on 24 February 2020. For us, Bürokratt was part of an overall attempt to really make the government embrace and adopt AI at large.

What Bürokratt added was that as opposed to just efficiency gains, we saw that it could be a way to radically improve government interactions, or how people interact with the government. It would make it more natural because we've had the issue of online services being hard to use for the elderly and others. Voice is the most natural thing, so we were really banking on the hopeful trend of AI virtual assistants becoming a major thing soon. And that is why we started experimenting, so if AI virtual assistants would become a thing, then our government would be ready. Bürokratt was really like an experimentation program and part of the overall AI movement [in the government].

Outside of the existing plans for the Bürokratt system, can you talk about what AI will enable the government to do and why it is potentially so powerful?

I have to go into the story in context a bit. For Estonia, the driving factor or motivation as to why to do digital was always productivity in the government. Being a small place as we are, no natural resources really and so forth… you have to be very efficient with the public funds that we have. This carried the wave of digitalization, but what started to happen was that in terms of productivity or efficiency gains, the marginal benefit became smaller and smaller in the next digital iterations. Thus, the effect was plateauing.

Now, AI could give a new exponential leap in that direction, was the hypothesis. Already the first pilot projects showed it really is so. Thus, for us, it was about unleashing the next level of productivity gain. I think the most exciting use cases initially were actually concrete low-hanging fruits. That had been our hypothesis to say, "hey, you don't have to talk about gen AI or whatever... even simple, narrow AI in a concrete machine learning application, if used wisely, can boost you." Our classic first use case example is having machines interpret satellite imagery to understand which fields on agricultural grants are currently being taken care of or not. In a nutshell, it's a very narrow, simple use case, but immediately paid off in a few months.

I think we've only scratched the surface in Estonia in many ways. We haven't really gotten deep into what can be done yet in that sense. I think the most exciting things will come in the second or third waves of maturity possibly, like doctors having a smart AI assistant as they go through diagnostics and prescriptions to recommend immediate actions. Basically, if they see you, it's not just that they have your medical history available, actually they already have targeted, personalized advice available as to what perhaps should have been offered to you. That sort of stuff. Technologically, it's so complex it's still a bit far off, but fundamentally that's the idea.

From the current use cases, I'm very excited about things that people notice. Which are things like having real-time parliament hearing minutes. Which for me was always crazy because you had even just a few years ago still actual people doing the typing up of meeting minutes, which our phones can basically do any minute. Things like that will show a noticeable next-level change, an improvement in something that actually is a silly thing that otherwise people have to do manually.

Can you talk a little bit more about how AI will shape the future of government? What about the future of the nation-state?

The nation-state is a trickier one, because that's a loaded notion: what's going to happen to notions of citizenship and things like that? Well, first, given everything we've said so far, I'm clearly very optimistic and biased. I can clearly imagine that there can be government who are hesitant, but AI is no different than the rest of digital transformation in many ways. It's another tool or another stage for transformation. Even today, there are governments who have stood out by fully adopting digital, embracing it, and those who have been really reluctant and hesitant. And then there's the middle ground, who perhaps are trying to but haven't gotten there yet. My point being, with AI, governments act the same. But what we also see in embracing digital writ large, those who do it reap certain developmental benefits in terms of investment or growth or also public trust.

My point is that as the technology is here: if you don't embrace it, you will lose out. If not for anything else, you'll lose out compared to what people expect from government services. You have this dichotomy that you're just not doing as good of a job as the government or you lose out competitively in the race with other governments, which exists especially in the business world anyway. That's my simplistic thinking to the question "what's the future of government?": if you don't go AI, you just start to lose.

ACKNOWLEDGEMENTS

Mom and Dad, your love and support made not just this book but so many things in my life possible. You are two of the kindest, most selfless, and thoughtful people in the world. I am so fortunate to be your son. Thank you. Dominic, I am proud of the man you are and proud to be your brother.

This book would not have been possible without two other people—Geoff Cain and Lacey Strahm. Geoff's sage advice at crucial moments helped turn this book from an idea to a reality. Lacey's feedback and passion pushed me to keep going in moments where I felt like giving up.

There are several people who supported me with their editorial advice who I must thank. Professor Ben Noble, who took a chance on a first-time writer and pushed me to create the best version of the book that I could. Alasdair Craig, whose many insights aided me in making the book that much more compelling. Will Murphy, who helped me turn an incredibly basic idea into a real outline and who shepherded me through the beginning stages of the writing process. Finally, Michael Dwyer and the entire Hurst team—your work supporting authors with diverse views is an inspiration and I am grateful to be a part of the Hurst family.

This book would not have been possible without the generous support of so many who gave their time, provided access to their research, and answered unsolicited questions from a stranger on the internet. My heartfelt thanks to Siim Sikkut, who not only helped to inspire this book but was incredibly generous with his time, sitting

ACKNOWLEDGEMENTS

for multiple interviews and providing valuable insights and feedback throughout the process. Thank you to President Toomas Hendrik Ilves, Prime Minister Laar, Sten Tamkivi, MP Hanah Lahe, Dr. Matthew Mitchell, Jennifer Pahlka, Marina Nitze, Steve Jurvetson, Professor Rainer Kattel, Dr. Arvo Ott, Dr. Keegan McBride, Jaan Tallinn, Dmitri Jegorov, Minister Andres Sutt, Paul Goble, Mike Bracken, Dr. Leslie Berlin, Dr. Piret Ehin, Ruth Annus, Veiko Lember, Aaron Maniam, Professor Wolfgang Drechsler, Andrus Viirg, Merle Maigre, Professor Pille Pruulmann-Vengerfeldt, Atossa Araxia Abrahamian, Arne Ansper and Anna Klimovits, Dennis Linders, Dr. Aet Annist, Dr. Samuli Haataja, Janek Metsallik, Pille Muni, Dr. Jaan Elias, Professor Donald Moynihan, Dr. Una Bergmane, Kalev Aasmae, Liisi Esse, Ott Velsberg, Ants Sild, Henrik Roonemaa, Dmitri Jegorov, Ruth Annus, Tom Kalil, Ott Kaukver and many others.

I must also acknowledge the innumerable scholars, journalists, and authors whose work I built on, especially Prime Minister Mart Laar, Professor Rainer Kattel and the broader UCL academic community, Dr. Una Bergmane, and Neil Taylor, as well as Dr. Matthew Mitchell and the authors of the *Road to Freedom*, the collected works of the e-Governance Academy, Liisi Esse at the Stanford Library which hosts a great number of excellent interviews with Estonian leaders and members of the broader Estonian community and diaspora, the International Centre for Defence and Security, and finally Silver Tambur whose English language Estonian World publication has afforded me many helpful insights into the Estonian nation.

P.S. If you've gotten this far, thank you for reading. I hope you've enjoyed the book—if you'd like to connect with me directly, my email is joelburke2014@gmail.com.

NOTES

INTRODUCTION

1. Innovator's Dilemma: the Innovator's Dilemma is a theory posited by Harvard professor Clayton Christensen which helps to explain why established and mature companies often miss new technological paradigms due to an inability to disrupt their current business model.

1. THE BIRTH OF A UNICORN

1. "Estonia, an early country to introduce Christmas trees, celebrates the holiday. See the photos," *AP News*, last modified 23 December 2023, https://apnews.com/article/estonia-christmas-trees-tallinn-baltic-solstice-lights-968cf0167f-d14ae9a05a8b96ae614c0f.
2. Hannah Boland, "Meet Bolt's Markus Villig, Europe's youngest founder of a billion pound company, who is taking on Uber," *The Telegraph*, 28 August 2019, https://www.telegraph.co.uk/technology/2019/07/14/meet-bolts-markus-villig-europes-youngest-founder-1-billion/.
3. "12 digital services in e-Estonia," Visit Estonia, https://www.visitestonia.com/en/why-estonia/12-digital-services-in-e-estonia.
4. "Estonia leads Europe in startups, unicorns, and investments per capita," Invest in Estonia, https://investinestonia.com/estonia-leads-europe-in-startups-unicorns-and-investments-per-capita/.
5. "Estonia No 1 in UN ranking of e-services, Tallinn 3rd," *The Baltic Times*, https://www.baltictimes.com/estonia_no_1_in_un_ranking_of_e-services__tallinn_3rd/.
6. Anna Bocharnikova, "Economic well-being under plan versus market: The case of Estonia and Finland," *CATO Journal*, Winter 2021, https://www.cato.org/cato-journal/winter-2021/economic-well-being-under-plan-versus-market-case-estonia-finland.

7. Matthew D. Mitchell, Peter J. Boettke, and Konstantin Zhukov, *The Road to Freedom* (Fraser Institute, 2023), 64, https://www.fraserinstitute.org/sites/default/files/road-to-freedom-estonias-rise-from-soviet-vassal-state-to-one-of-the-freest-nations-on-earth_0.pdf.
8. Kristie Pickering, "The digital nomad's guide to Estonia," *Sifted*, 29 June 2023, https://sifted.eu/articles/digital-nomads-guide-estonia-brnd.
9. "History," History & Index, Bluemoon, http://www.bluemoon.ee/history/index.html.
10. The game is still playable on the Internet Archive and is an excellent nostalgia trip.
11. Silver Tambur, "Skype and the Estonian start-up ecosystem," *Estonian World*, 1 September 2013, https://estonianworld.com/technology/skype-estonian-start-ecosystem/.
12. Matthew D. Mitchell, Peter J. Boettke, and Konstantin Zhukov, *The Road to Freedom* (Fraser Institute, 2023), 126, https://www.fraserinstitute.org/sites/default/files/road-to-freedom-estonias-rise-from-soviet-vassal-state-to-one-of-the-freest-nations-on-earth_0.pdf.
13. "A conversation with Jaan Tallinn," *Edge*, 16 April 2015, https://www.edge.org/conversation/jaan_tallinn-existential-risk.
14. Toivo Tänavsuu, "'How can they be so good?': The strange story of Skype," *Ars Technica*, 3 September 2018, https://arstechnica.com/information-technology/2018/09/skypes-secrets/.
15. Jonathan Krim, "Kazaa co-founder has new calling," *NBC News*, 4 June 2005, https://www.nbcnews.com/id/wbna8091539.
16. "Everyday.com portal launched in Spain," Cision, https://news.cision.com/mtg/r/everyday-com-portal-launched-in-spain,c35755.
17. William L. Hosch, "Janus Friis," *Britannica*, https://www.britannica.com/biography/Janus-Friis.
18. Jonathan Krim, "Kazaa co-founder has new calling," *NBC News*, 4 June 2005, https://www.nbcnews.com/id/wbna8091539.
19. "1999 income data, revised September 2001," Enforcement and Compliance, Trade.gov, https://enforcement.trade.gov/wages/99wages/99wages.htm.
20. Toivo Tänavsuu, "'How can they be so good?': The strange story of Skype," *Ars Technica*, 3 September 2018, https://arstechnica.com/information-technology/2018/09/skypes-secrets/.
21. David Rowan, "What I've learned by Skype's Niklas Zennstrom," *Wired*, 6 November 2010, https://www.wired.com/2010/11/what-ive-learned-by-skypes-niklas-zennstrom/.
22. Toivo Tänavsuu, "'How can they be so good?': The strange story of Skype," *Ars Technica*, 3 September 2018, https://arstechnica.com/information-technology/2018/09/skypes-secrets/.
23. Ben Gilbert and David Rosenthal, "Skype," Season 1, Episode 24, https://www.acquired.fm/episodes/episode-24-skype.

24. "Kazaa," Encylopedia, *PC Mag*, https://www.pcmag.com/encyclopedia/term/kazaa.
25. Jonathan Krim, "Kazaa co-founder has new calling," *NBC News*, 4 June 2005, https://www.nbcnews.com/id/wbna8091539.
26. Jan Libbenga, "Dutch Supreme Court rules Kazaa legal," *The Register*, 19 December 2003, https://www.theregister.com/2003/12/19/dutch_supreme_court_rules_kazaa/.
27. "Dutch court throws out Kazaa case," *Wired*, 19 December 2003, https://www.wired.com/2003/12/dutch-court-throws-out-kazaa-case/.
28. Michele Chandler, "European high-tech startups thrive, says Skype founder," *Insights by Stanford* Business, 1 March 2011, https://www.gsb.stanford.edu/insights/european-high-tech-startups-thrive-says-skype-founder.
29. Toivo Tänavsuu, "'How can they be so good?': The strange story of Skype," *Ars Technica*, 3 September 2018, https://arstechnica.com/information-technology/2018/09/skypes-secrets/.
30. Jonathan Krim, "Kazaa co-founder has new calling," *NBC News*, 4 June 2005, https://www.nbcnews.com/id/wbna8091539.
31. Toivo Tänavsuu, "'How can they be so good?': The strange story of Skype," *Ars Technica*, 3 September 2018, https://arstechnica.com/information-technology/2018/09/skypes-secrets/.
32. Toivo Tänavsuu, "'How can they be so good?': The strange story of Skype," *Ars Technica*, 3 September 2018, https://arstechnica.com/information-technology/2018/09/skypes-secrets/.
33. "Kazaa Sells Site, Software," *Wired*, 21 January 2002, https://www.wired.com/2002/01/kazaa-sells-site-software/.
34. Toivo Tänavsuu, "'How can they be so good?': The strange story of Skype," *Ars Technica*, 3 September 2018, https://arstechnica.com/information-technology/2018/09/skypes-secrets/.
35. Sten Tamkivi (@seikatsu), "True classic: the original Skype product discussion memo from Toivo (RIP) to @nzennstrom & @jaanusfriis (dug up by Hannu to commemorate 20 years from launch)," Twitter photo, 29 August 2023, https://twitter.com/seikatsu/status/1696488438755684587?s=46&t=IzPrVbo2SjIuAkI3DMvhxw.
36. "FCC releases statistics of the long distance telecommunications industry report," News, Federal Communications Commission, 14 May 2003, https://transition.fcc.gov/Bureaus/Common_Carrier/Reports/FCC-State_Link/IAD/ldrpt103.pdf.
37. Kara Swisher, *Burn Book* (New York: Simon & Schuster 2024), 120.
38. Toivo Tänavsuu, "'How can they be so good?': The strange story of Skype," *Ars Technica*, 3 September 2018, https://arstechnica.com/information-technology/2018/09/skypes-secrets/.
39. "Skype," Science and Tech, *Britannica*, https://www.britannica.com/technology/Skype.

40. Ben Gilbert and David Rosenthal, "Skype," Season 1, Episode 24, https://www.acquired.fm/episodes/episode-24-skype.
41. Ben Gilbert and David Rosenthal, "Skype," Season 1, Episode 24, https://www.acquired.fm/episodes/episode-24-skype.
42. Michele Chandler, "European high-tech startups thrive, says Skype founder," *Insights by Stanford Business*, 1 March 2011, https://www.gsb.stanford.edu/insights/european-high-tech-startups-thrive-says-skype-founder.
43. David Rowan, "What I've learned by Skype's Niklas Zennstrom," *Wired*, 6 November 2010, https://www.wired.com/2010/11/what-ive-learned-by-skypes-niklas-zennstrom/.
44. Ferdinand v. Götzen, "An interview with Jaan Tallinn, co-founder and author of Skype," https://ferdinandgoetzen.wordpress.com/2014/12/01/an-interview-with-jaan-tallinn-co-founder-and-author-of-skype/.
45. Kevin J. O'Brien, "Microsoft inherits sticky data collection issues from Skype," *The New York Times*, 24 February 2013, https://www.nytimes.com/2013/02/25/technology/microsoft-inherits-sticky-data-collection-issues-from-skype.html.
46. Berlin, Leslie, "The first venture capital firm in Silicon Valley: Draper, Gaither & Anderson," in Bruce J. Schulman (ed.), *Making the American Century: Essays on the Political Culture of Twentieth Century America* (New York, 2014; online edn, Oxford Academic, 16 April 2014), https://doi.org/10.1093/acprof:oso/9780199845392.003.0010.
47. "The Draper family, an unparalleled legacy in venture capital," Draper, Draper Cygnus, https://drapercygnus.vc/draper/.
48. "The Draper family, an unparalleled legacy in venture capital," Draper, Draper Cygnus, https://drapercygnus.vc/draper/.
49. "Meet the startups making sci-fi a reality," Portfolio, BoostVC, tpps://www.boost.vc/portfolio.
50. Ari Levy, "Coinbase closes at $328.28 per share in Nasdaq debut, valuing crypto exchange at $85.8 billion," *CNBC*, https://www.cnbc.com/2021/04/14/coinbase-to-debut-on-nasdaq-in-direct-listing.html.
51. Brad Meikle, "Draper cashes in on billion-dollar Skype," 19 September 2005, https://www.buyoutsinsider.com/draper-cashes-in-on-billion-dollar-skype/.
52. Chris O'Brien, "Shedding Soviet history, Estonia aims to be world's most pro-tech nation," *VentureBeat*, 25 February 2017, https://venturebeat.com/business/estonia-digital-nation/.
53. Dan Scott and KG, "The final frontier with Steve Jurvetson: Part one," *Cloud Valley*, 10 September 2022, https://cloudvalley.substack.com/p/the-final-frontier-with-steve-jurvetson.
54. Toivo Tänavsuu, "'How can they be so good?': The strange story of Skype," *Ars Technica*, 3 September 2018, https://arstechnica.com/information-technology/2018/09/skypes-secrets/.
55. Michael Kanellos, "The big winner behind Skype and Baidu," *CNET*, 15 September

2005, https://www.cnet.com/tech/tech-industry/the-big-winner-behind-skype-and-baidu/.

56. Toivo Tänavsuu, "'How can they be so good?': The strange story of Skype," *Ars Technica*, 3 September 2018, https://arstechnica.com/information-technology/2018/09/skypes-secrets/.

57. Toivo Tänavsuu, "'How can they be so good?': The strange story of Skype," *Ars Technica*, 3 September 2018, https://arstechnica.com/information-technology/2018/09/skypes-secrets/.

58. "eBay Inc. reiterates 'the truth about Skype,'" Press Room, eBay, 3 March 2014, https://www.ebayinc.com/stories/news/ebay-inc-reiterates-truth-about-skype/.

59. Ott Kaukver, "Oral history interview with Ott Kaukver," *Baltic Video Testimonies at Stanford Libraries*, 26 August 2017, video, 26:48, https://purl.stanford.edu/hv021wq1614.

60. Sten Tamkivi (@seikatsu), "This weekend a few hundred early Skypers gathered in Tallinn to celebrate 20 years from the first beta release of Skype, on 23 August 2003. #skype20. Here's a thread on how this company left a dent," Twitter thread, 18 September 2023, https://twitter.com/seikatsu/status/1703676016395231383.

61. Sam Shead, "I invest in AI. It's the biggest risk to humanity," *Forbes*, 21 August 2019, https://www.forbes.com/sites/samshead/2019/08/21/the-skype-mafia-who-are-they-and-where-are-they-now/?sh=ebef2fa73994.

62. "Wise," Organization, Crunchbase, https://www.crunchbase.com/organization/transferwise.

63. Rey Mashayekhi, "Wise co-founder Taavet Hinrikus on his next act as a venture capitalist," *Fortune*, 24 November 2021, https://fortune.com/2021/11/24/wise-taavet-hinrikus-venture-capital-tech-startup-investing/.

64. Amy Lewin, "Taavet Hinrikus and Ian Hogarth launch € 250m VC fund," *Sifted*, 27 June 2022, https://sifted.eu/articles/taavet-hinrikus-vc-fund-plural.

65. Craig Turp-Balazs, "Why Skype remains key to Estonia's digital success," *Emerging Europe*, 15 April 2020, https://emerging-europe.com/business/why-skype-remains-key-to-estonias-digital-success/.

66. Silver Tambur, "Skype and the Estonian start-up ecosystem," 1 September 2013, https://estonianworld.com/technology/skype-estonian-start-ecosystem/.

67. "A Conversation With Jaan Tallinn," *Edge*, 16 April 2015, https://www.edge.org/conversation/jaan_tallinn-existential-risk.

68. Jaan Tallinn, "I invest in AI. It's the biggest risk to humanity," *Newsweek*, 4 May 2023, https://www.newsweek.com/artificial-intelligence-tech-risk-skype-google-1798280.

69. Alexandra Wolfe, "Jaan Tallinn," *The Wall Street Journal*, 31 May 2013, https://www.lesswrong.com/posts/JrsoPnKK5X3zwSgj7/jaan-tallinn-a-skype-founder-on-biomonitors-existential-risk.

70. Mara Hvistendahl, "Can we stop AI outsmarting humanity?," *The Guardian*,

28 March 2019, https://www.theguardian.com/technology/2019/mar/28/can-we-stop-robots-outsmarting-humanity-artificial-intelligence-singularity.
71. "Pause giant AI experiments: An open letter," Open Letter, Future of Life Institute, https://futureoflife.org/open-letter/pause-giant-ai-experiments/.

2. SONIC YOUTH

1. "Ei ole üksi ükski maa," Estonian patriotic song, Wikiwand, https://www.wikiwand.com/en/Ei_ole_%C3%BCksi_%C3%BCkski_maa.
2. Toivo U. Raun, *Estonia and the Estonians*, second edition, updated (Stanford: Hoover Institution Press, 2002), Kindle edition, location 3123.
3. Mehmet Oğuzhan Tulun, "The Baltic people by the Russian Empire and the Soviet Union," Başkent University, 25 January 2012, https://dergipark.org.tr/tr/download/article-file/701004.
4. Olivia B. Waxman, "Mikhail Gorbachev championed 'glasnost' and 'perestroika.' Here's how they changed the world," *Time*, 30 August 2022, https://time.com/5512665/mikhail-gorbachev-glasnost-perestroika/.
5. Kaja Kallas: Member of Parliament 2011–2014 and 2019–2021, Prime Minister 2021–2024.
6. Kaja Kallas, "How Estonia became a model for digital democracy," interview by Azeem Azhar, *Exponentially with Azeem Azhar*, Bloomberg, 7 September 2023, video, 3:59, https://www.bloomberg.com/news/articles/2023-09-07/video-how-estonia-became-a-digital-democracy-and-avoided-toxic-social-media.
7. For those interested in learning more about Estonian and Baltic resistance, *The Forest Brotherhood: Baltic Resistance against the Nazis and Soviets* by Dan Kaszeta presents an excellent in-depth look.
8. Toivo U. Raun, *Estonia and the Estonians*, second edition, updated (Stanford: Hoover Institution Press, 2002), Kindle edition, location 2150.
9. Marju Lauristin: Minister of Social Affairs 1992–1994, Member of Estonian Parliament 1992–2014, Member of the European Parliament, 2014–2017.
10. Anthony Lewis, "ABROAD AT HOME: The Estonian test," *The New York Times*, 2 November 1989, https://www.nytimes.com/1989/11/02/opinion/abroad-at-home-the-estonian-test.html.
11. Matthew D. Mitchell, Peter J. Boettke, and Konstantin Zhukov, *The Road to Freedom* (Fraser Institute, 2023), 97–98, https://www.fraserinstitute.org/sites/default/files/road-to-freedom-estonias-rise-from-soviet-vassal-state-to-one-of-the-freest-nations-on-earth_0.pdf.
12. John Moody, "Gorbachev loosens the reins," *Time*, 30 June 1986, https://time.com/vault/issue/1986-06-30/page/55/.
13. Neil Taylor, *Estonia: A Modern History* (London: C. Hurst & Co., 2018), Kindle edition, 152.
14. Tiit Made: A former politician and member of the "August 20 Club" formed of representatives who voted for the restoration of the independence of Estonia. Author of *Estonians' Liberation Way*.

NOTES

15. Tiit Made, *Estonians' Liberation Way* (August 20 Club, 2015), 20.
16. "Estonians stop toxic phosphorite mining," 1987–88, Global Nonviolent Action Database, https://nvdatabase.swarthmore.edu/content/estonians-stop-toxic-phosphorite-mining-1987-88.
17. Anatol Lieven, *The Baltic Revolution: Estonia, Latvia, Lithuania and the Path to Independence* (New Haven and London, Yale University Press, 1993), 220.
18. Mart Laar: Prime Minister 1992–1994 and 1999–2002, Minister of Defense 2011–2012.
19. Mart Laar, "Estonia—The small country that could," *Institute for Democracy in Eastern Europe*, https://idee.org/cubalaar.html.
20. Siim Kallas: Member of the Supreme Council of the Soviet Union 1989–1991, Member of Parliament of the Republic of Estonia 1995–2004, Minister of Foreign Affairs 1995–1996, Minister of Finance 1999–2002, Prime Minister 2003, EU Commissioner 2004–2010. Father of Prime Minister Kaja Kallas.
21. Edgar Savisaar: Prime Minister of the Interim Government of Estonia 1991–1992, Minister of Internal Affairs 1995, Minister of Economic Affairs and Communications 2005–2007.
22. "Upheaval in the East; 100 leading Estonians ask for independence," *The New York Times*, https://www.nytimes.com/1990/02/09/world/upheaval-in-the-east-100-leading-estonians-ask-independence.html.
23. Tiit Made, *Estonians' Liberation Way* (August 20 Club, 2015), 30.
24. *The Singing Revolution*. Directed by Maureen Tusty, performances by James Tusty, Docurama, 2006. Time: 22:15.
25. Anatol Lieven, *The Baltic Revolution: Estonia, Latvia, Lithuania and the Path to Independence* (New Haven and London, Yale University Press, 1993), 110.
26. Katharine Schwab, "A country created through music," *The Atlantic*, 12 November 2015, https://www.theatlantic.com/international/archive/2015/11/estonia-music-singing-revolution/415464/.
27. Arvo Pärt: an Estonian-born composer of contemporary classical music. From 2011 to 2018 Pärt was the most performed living composer in the world.
28. "Arvo Pärt," Biography, Arvo Pärt Centre, https://www.arvopart.ee/en/arvo-part/.
29. Sandra Saar, "Song and Dance Festival ignites deep sense of Estonian identity," *ERR*, 29 June 2023, https://news.err.ee/1609020947/song-and-dance-festival-ignites-deep-sense-of-estonian-identity#:~:text=The%20general%20Estonian%20Song%20and,the%20right%20decision%20in%20hindsight.
30. Toivo U. Raun, *Estonia and the Estonians*, second edition, updated (Stanford: Hoover Institution Press, 2002), Kindle edition, Location 3130.
31. Stephen Zunes, "Estonia's Singing Revolution (1986–1991)," International Center on Nonviolent Conflict, April 2009, https://www.nonviolent-conflict.org/estonias-singing-revolution-1986-1991/#:~:text=In%20September%201988%2C%20a%20massive,for%20restoring%20the%20country's%20independence.

32. Tiit Made, *Estonians' Liberation Way* (August 20 Club, 2015), 49–50.
33. Tiit Made, *Estonians' Liberation Way* (August 20 Club, 2015), 49–50.
34. *The Singing Revolution*. Directed by Maureen Tusty, performances by James Tusty, Docurama, 2006, Time: 46:57.
35. Guntis Smidchens, *The Power of Song: Nonviolent National Culture in the Baltic Singing Revolution* (Seattle and London: University of Washington Press, 2013), 252.
36. Tiit Made, *Estonians' Liberation Way* (August 20 Club, 2015), 64.
37. Tiit Made, *Estonians' Liberation Way* (August 20 Club, 2015), 63–64.
38. Tiit Made, *Estonians' Liberation Way* (August 20 Club, 2015), 63–64.
39. Tiit Made, *Estonians' Liberation Way* (August 20 Club, 2015), 81.
40. Professor Archie Brown, "Context, cross-pressures and compromise: The roles of Gorbachev and Yeltsin," *Understanding the Baltic States: Estonia, Latvia and Lithuania since 1991*, ed. Charles Clarke (London: C. Hurst & Co, 2023), Kindle edition, 141.
41. Mart Laar, "The Estonian economic miracle," *The Heritage Foundation*, 7 August 2007, https://www.heritage.org/report/the-estonian-economic-miracle.
42. David, Remnick, "Kremlin acknowledges secret pact on Baltics; Soviets deny republics annexed illegally," *The Washington Post*, 19 August 1989, https://www.nytimes.com/1989/08/19/world/soviets-confirm-nazi-pacts-dividing-europe.html.
43. Tiit Made, *Estonians' Liberation Way* (August 20 Club, 2015), 81.
44. Lucian Kim, "The Human Chain of the Baltic Way," *NPR*, 24 August 2019, https://www.npr.org/2019/08/24/753962667/the-human-chain-of-the-baltic-way.
45. Helen Wright and Silver Tambur, "The Baltic Way—The longest unbroken human chain in history," *Estonian World*, 23 August 2021, https://estonianworld.com/life/estonia-commemorates-30-years-since-the-baltic-way-the-longest-unbroken-human-chain-in-history/.

3. INDEPENDENCE DAY

1. Tiit Made, *Estonians' Liberation Way* (August 20 Club, 2015), 90.
2. Martha Brill Olcott, "The Lithuania crisis," *Foreign Affairs*, 1 June 1990, https://www.foreignaffairs.com/articles/russian-federation/1990-06-01/lithuanian-crisis.
3. Audrius Siaurasevicius and John Rettie, "Lithuania breaks away from the Soviet Union," *The Guardian*, 12 March 1990, https://www.theguardian.com/world/1990/mar/12/eu.politics.
4. Doug Bandow, "How Lithuania destroyed the Soviet Union," *CATO Institute*, 7 June 2019, https://www.cato.org/commentary/how-lithuania-destroyed-soviet-union.
5. Una Bergmane, *Politics of Uncertainty: The United States, the Baltic Question,*

and the Collapse of the Soviet Union (New York, NY: Oxford University Press, 2023), 79.

6. Audrius Siaurusevicius, "Soviet tanks crush the human shield of Vilnius," *The Guardian*, 14 January 1991, https://www.theguardian.com/world/1991/jan/14/eu.politics.
7. Michael Dobbs, "Soviet troops seize Lithuania's TV station," *The Washington Post*, 12 January 1991, https://www.washingtonpost.com/archive/politics/1991/01/13/soviet-troops-seize-lithuanias-tv-station/34739858-1b02-4f92-85e8-ab612554fd4e/.
8. Marek Grzegorczyk, "Lithuania remembers January 13, 1991," *Emerging Europe*, 13 January 2021, https://emerging-europe.com/news/lithuania-remembers-january-13-1991/.
9. Una Bergmane, *Politics of Uncertainty: The United States, the Baltic Question, and the Collapse of the Soviet Union* (New York, NY: Oxford University Press, 2023), 118.
10. "Timeline: Former Russian President Boris Yeltsin," World, *NPR*, https://www.npr.org/2007/04/23/9774006/timeline-former-russian-president-boris-yeltsin.
11. Paul Goble, "Vilnius at 30—Nothing must be forgotten," *Eurasia Daily Monitor*, vol. 18, no. 6 (12 January 2021), https://jamestown.org/program/vilnius-at-30-nothing-must-be-forgotten/.
12. Serge Schmemann, "7 Soviet republics agree to seek new union," *The New York Times*, 15 November 1991, https://www.nytimes.com/1991/11/15/world/7-soviet-republics-agree-to-seek-new-union.html.
13. David Remnick, "Gorbachev unveils his new union treaty," *The Washington Post*, 23 November 1990, https://www.washingtonpost.com/archive/politics/1990/11/24/gorbachev-unveils-his-new-union-treaty/44a51b57-c258-4d29-a0ff-625b31bd0d1d/.
14. David Remnick, "Gorbachev unveils his new union treaty," *The Washington Post*, 23 November 1990, https://www.washingtonpost.com/archive/politics/1990/11/24/gorbachev-unveils-his-new-union-treaty/44a51b57-c258-4d29-a0ff-625b31bd0d1d/.
15. Charles Maynes, "In 1991, Soviet citizens saw swans on the TV... And knew it meant turmoil," *NPR*, 19 August 2021, https://www.npr.org/2021/08/19/1029437787/in-1991-soviet-citizens-saw-swans-on-the-tv-and-knew-it-meant-turmoil.
16. "Q+A: What was the hardline Soviet coup attempt in 1991?," World, *Reuters*, https://www.reuters.com/article/us-russia-coup-qa/qa-what-was-the-hardline-soviet-coup-attempt-in-1991-idUSTRE77F2GC20110816.
17. "Yeltsin's speech," Vancouver Island University, https://web.viu.ca/davies/H102/Yelstin.speech.1991.htm.
18. Michael Dobbs, "The 1991 coup attempt that the Soviet president barely survived," *The Washington Post*, https://www.washingtonpost.com/history/2022/08/30/coup-attempt-mikhail-gorbachev/.

19. Vladimir Isachenkov, "Hardline coup set the stage for Soviet collapse 30 years ago," *AP*, 18 August 2021, https://apnews.com/article/europe-fca50dc443b-66b53a67fc8043b9c2d2a.
20. Walter R. Iwaskiw (ed.), "Estonia, Latvia, and Lithuania country studies," Federal Research Division of the Library of Congress (January 1995), 24.
21. Einar Vära, "Estonia celebrates the restoration of independence," *Estonian World*, 20 August 2023, https://estonianworld.com/life/estonia-celebrates-the-day-of-restoration-of-independence/.
22. Una Bergmane, *Politics of Uncertainty: The United States, the Baltic Question, and the Collapse of the Soviet Union* (New York, NY: Oxford University Press, 2023), 130.
23. "Twenty-eight years since Estonia regained independence from Soviet Union," *ERR*, 20 August 2019, https://news.err.ee/972006/twenty-eight-years-since-estonia-regained-independence-from-soviet-union.
24. Masha Gessen, "Why Estonia was poised to handle how a pandemic would change everything," *The New Yorker*, 24 March 2020, https://www.newyorker.com/news/our-columnists/why-estonia-was-poised-to-handle-how-a-pandemic-would-change-everything.
25. "The defenders of the TV tower," Huvasti Charlie, https://virtuaalne.ajaloomuuseum.ee/huvasticharlie/index087f.html?id=10755.
26. Tiit Made, *Estonians' Liberation Way* (August 20 Club, 2015), 139.
27. Una Bergmane, *Politics of Uncertainty: The United States, the Baltic Question, and the Collapse of the Soviet Union* (New York, NY: Oxford University Press, 2023), 121.
28. Una Bergmane, *Politics of Uncertainty: The United States, the Baltic Question, and the Collapse of the Soviet Union* (New York, NY: Oxford University Press, 2023), 154.
29. "The defenders of the TV tower," Huvasti Charlie, https://virtuaalne.ajaloomuuseum.ee/huvasticharlie/index087f.html?id=10755.
30. Paavo Palk, "Revelations on the restoration of independence," *RKK ICDS*, 19 August 2016, https://icds.ee/en/revelations-on-the-restoration-of-independence/.
31. Paavo Palk, "Revelations on the restoration of independence," *RKK ICDS*, 19 August 2016, https://icds.ee/en/revelations-on-the-restoration-of-independence/.

4. A NATION REFORMED

1. Jako Salla, Vania Ceccato, and Andri Ahven, "Homicide in Estonia," in Marieke C. A. Liem, William Alex Pridemore (eds), *Handbook of European Homicide Research*, New York, NY: 2012), https://doi.org/10.1007/978-1-4614-0466-8_27.
2. "Homicide rates over the long term," Homicide Rates across Western Europe, Our World in Data, https://ourworldindata.org/grapher/homicide-rates-across-western-europe?tab=table&time=1994..latest.

NOTES

3. "Mexico's homicide rate dropped in 2022, but appears to flatline in 2023, official figures show," *AP News*, 25 July 2023, https://apnews.com/article/mexico-homicides-rate-violence-d0a9a83c3124b1f3ce0e8af6c9b8aa9f.
4. Walter R. Iwaskiw (ed.), "Estonia, Latvia, and Lithuania country studies," Federal Research Division of the Library of Congress (January 1995), 78.
5. Joshua J. Mark, "Hanseatic League," *World History Encyclopedia*, 8 March 2019, https://www.worldhistory.org/Hanseatic_League/.
6. Anatol Lieven, *The Baltic Revolution: Estonia, Latvia, Lithuania and the Path to Independence* (New Haven and London, Yale University Press, 1993), 317.
7. Mark Galeotti, "Feature: Beware the underworld merchant-adventurer," *ERR*, 16 May 2019, https://news.err.ee/940477/feature-beware-the-underworld-merchant-adventurer.
8. Mark Galeotti, "Feature: Beware the underworld merchant-adventurer," *ERR*, 16 May 2019, https://news.err.ee/940477/feature-beware-the-underworld-merchant-adventurer.
9. *Rodeo—Taming a Wild Country*, directed by Raimo Jõerand, performances by Kiur Aarma, Kinocompany, 2018, time 20:20.
10. Lagle Parek: Minister of the Interior October 1992–November 1993. Parek was arrested in 1983 and sentenced to six years in prison and three in exile. After being released in 1987, Parek returned to Estonia and became one of the founders of the Estonian National Independence Party.
11. *Rodeo—Taming a Wild Country*, directed by Raimo Jõerand, performances by Kiur Aarma, Kinocompany, 2018, time 18:32.
12. Walter R. Iwaskiw (ed.), "Estonia, Latvia, and Lithuania country studies," Federal Research Division of the Library of Congress (January 1995), 78.
13. Anatol Lieven, *The Baltic Revolution: Estonia, Latvia, Lithuania and the Path to Independence* (New Haven and London, Yale University Press, 1993), 325.
14. Anatol Lieven, *The Baltic Revolution: Estonia, Latvia, Lithuania and the Path to Independence* (New Haven and London, Yale University Press, 1993), 325.
15. Mart Laar, "The Estonian economic miracle," *The Heritage Foundation*, 7 August 2007, https://www.heritage.org/report/the-estonian-economic-miracle.
16. Anatol Lieven, *The Baltic Revolution: Estonia, Latvia, Lithuania and the Path to Independence* (New Haven and London, Yale University Press, 1993), 357.
17. Tiit Vähi was first elected as a caretaker Prime Minister in January 1992 to October 1992, tasked with carrying out monetary reform. He would eventually serve as Prime Minister again from 1995 to 1997.
18. Anatol Lieven, *The Baltic Revolution: Estonia, Latvia, Lithuania and the Path to Independence* (New Haven and London, Yale University Press, 1993), 355–356.
19. Anatol Lieven, *The Baltic Revolution: Estonia, Latvia, Lithuania and the Path to Independence* (New Haven and London, Yale University Press, 1993), 202–203.
20. Jacques de Larosière and Steve H. Hanke, "A solution to Lebanon's economic crisis," *CATO Institute*, 23 September 2021, https://www.cato.org/commentary/solution-lebanons-economic-crisis.

21. Anatol Lieven, *The Baltic Revolution: Estonia, Latvia, Lithuania and the Path to Independence* (New Haven and London, Yale University Press, 1993), 347.
22. David Boaz, "The wisdom of the mart," *The Guardian*, 20 April 2006, https://www.theguardian.com/commentisfree/2006/apr/20/upfromcommunism.
23. "Mart Laar receives Milton Friedman Prize," Policy Report, CATO Institute, July/August 2006, https://www.cato.org/policy-report/july/august-2006/mart-laar-receives-milton-friedman-prize.
24. "Mart Laar receives Milton Friedman Prize," Policy Report, CATO Institute, July/August 2006, https://www.cato.org/policy-report/july/august-2006/mart-laar-receives-milton-friedman-prize.
25. Mart Laar, "The Estonian economic miracle," *The Heritage Foundation*, 7 August 2007, https://www.heritage.org/report/the-estonian-economic-miracle.
26. Rainer Kattel and Ringa Raudla, "Estonia's radical transformation: Successes and failures of 'crazy ideas,'" *The Economy 2030 Inquiry*, 18 November 2022, https://economy2030.resolutionfoundation.org/reports/estonias-radical-transformation/.
27. Author interview: Steve Jurvetson, email correspondence, 13 May 2024.
28. *History of the Digital State of Estonia: Formation of the Digital State 1991–2016*, commissioned by the Ministry of Economic Affairs and Communications and carried out by Ernst & Young, 6, https://www.ria.ee/media/1073/download.
29. Neil Taylor, *Estonia: A Modern History* (London: C. Hurst & Co., 2018), Kindle edition, 185.
30. Armen Sarkissian, *The Small States Club: How Small Start States Can Save the World* (London: C. Hurst & Co.), Kindle edition, 153.
31. "Estonia: Taxation system and implementation of flat income tax," Law Library, Library of Congress, December 2006, https://www.google.com/url?sa=t&source=web&rct=j&opi=89978449&url=https://tile.loc.gov/storage-services/service/ll/llglrd/2019669497/2019669497.pdf&ved=2ahUKEwit1JiN9J6FAxX2ElkFHbDkBm8QFnoECA4QAw&usg=AOvVaw2HW0UHwQ0I-uNLWLX1noaS.
32. *Rodeo—Taming a Wild Country*, directed by Raimo Jõerand, performances by Kiur Aarma, Kinocompany, 2018, time 57:07.
33. M. Laar, "Leading a successful transition: The Estonian miracle," *European View*, vol. 7, no. 1 (2008), 67–74, https://doi.org/10.1007/s12290-008-0024-z.
34. M. Laar, "Leading a successful transition: The Estonian miracle," *European View*, vol. 7, no. 1 (2008), 67–74, https://doi.org/10.1007/s12290-008-0024-z.
35. Kaja Kallas, "In Conversation with Prime Minister Kaja Kallas," interview by Terry with student questions, The Tallinn University Podcast, March 2023, audio, 14:17.
36. Kaja Kallas, "In Conversation with Prime Minister Kaja Kallas," interview by Terry with student questions, The Tallinn University Podcast, March 2023, audio, 14:17, https://open.spotify.com/episode/00fCay9CU600RA66d66U5d.

37. "Estonia GDP (current US$)," The World Bank, https://data.worldbank.org/indicator/NY.GDP.MKTP.CD?locations=EE.
38. "Average annual wages in Estonia from 1995 to 2020 (in euros)," Statista, https://www.statista.com/statistics/1262792/estonia-average-annual-wages/.
39. Sten Hankewitz, "Estonia's average salary in Q4 2023 €1,904," *Estonian World*, 7 March 2024, https://estonianworld.com/business/estonias-average-salary-in-q4-2023-e1904/.
40. "Life expectancy at birth," OECD Data, https://data.oecd.org/healthstat/life-expectancy-at-birth.htm.
41. Silver Tambur, "Estonia one of the most unequal countries in the eurozone," *Estonian World*, 1 June 2023, https://estonianworld.com/business/estonia-one-of-the-most-unequal-countries-in-the-eurozone/#:~:text=Five%20per%20cent%20of%20the,hold%20the%20rest%20between%20them.
42. Toivo U. Raun, "Estonia in the 1990s," *Journal of Baltic Studies*, vol. 32, no. 1 (2001), 19–43, https://doi.org/10.1080/01629770000000221.
43. Kersti Kaljulaid: President of Estonia from October 2016–October 2021. First female head of state since independence in 1918 and the youngest President ever at the time of her election.
44. Stefan Fina, Bastian Heider, and Märt Masso, "Unequal Estonia: Regional socio-economic disparities in Estonia," *Europa*, 2021, https://library.fes.de/pdf-files/bueros/baltikum/18551.pdf.
45. M. Laar, "Leading a successful transition: The Estonian miracle," *European View*, vol. 7, no. 1 (2008), 67–74, https://doi.org/10.1007/s12290-008-0024-z.
46. Walter R. Iwaskiw (ed.), "Estonia, Latvia, and Lithuania country studies," Federal Research Division of the Library of Congress (January 1995).
47. Neil Taylor, *Estonia: A Modern History* (London: C. Hurst & Co., 2018), Kindle edition, 168–169.
48. Neil Taylor, *Estonia: A Modern History* (London: C. Hurst & Co., 2018), Kindle edition, 168–169.
49. "Banking & financing," Invest in Estonia, https://investinestonia.com/business-in-estonia/financing/eu-and-national-support/.
50. Peeter Vihma, *Twenty Years of Building Digital Societies: Thinking about the Past and Future of Digital Transformation* (Tallinn: e-Governance Academy, 2023).
51. Jennifer Pahlka, "Culture eats policy," *Niskanen Center*, 21 June 2023, https://www.niskanencenter.org/culture-eats-policy/.
52. Lennart Meri: Minister of Foreign Affairs April 1990–March 1992, President October 1992–October 2001.
53. William Branigin, "Priit Vesilind, writer who penetrated the Iron Curtain, dies at 80," *The Washington Post*, 22 November 2023, https://www.washingtonpost.com/obituaries/2023/11/22/priit-vesilind-national-geographic-estonia-obit-died/.
54. Priit Vesilind, "A reflection of one's land," in Lennart Meri, *The Pathfinder*, Tiit Pruuli (ed.) (Tallinn, Varrak, 2009,) 116.

55. Riina Luik, "Why President Meri held press conference in airport toilet," *Postimees*, 30 October 2015, https://news.postimees.ee/3381693/why-president-meri-held-press-conference-in-airport-toilet.
56. "Gallery: Lennart Meri at 90," *ERR*, 29 March 2019, https://news.err.ee/925076/gallery-lennart-meri-at-90.
57. Toivo U. Raun "Estonia in the 1990s," *Journal of Baltic Studies*, vol. 32 (2001).
58. Heido Vitsur, "A hundred years of the Estonian economy," *ERR*, 19 August 2021, https://estonianworld.com/business/a-hundred-years-of-the-estonian-economy/.
59. M. Laar, "Leading a successful transition: The Estonian miracle," *European View*, vol. 7, no. 1 (2008), 67–74, https://doi.org/10.1007/s12290-008-0024-z.
60. *Rodeo—Taming a Wild Country*, directed by Raimo Jõerand, performances by Kiur Aarma, Kinocompany, 2018, time 16:37.
61. Lee Kuan Yew, *From Third World to First: The Singapore Story: 1965–2000* (New York, NY: HarperCollins Publishers, 2000), 68.
62. *Rodeo—Taming a Wild Country*, directed by Raimo Jõerand, performances by Kiur Aarma, Kinocompany, 2018, time 17:02.
63. Toomas Hendrik Ilves: Ambassador to the United States from 1993–1996, Minister of Foreign Affairs December 1996–September 1998, President October 2006–October 2016.

5. PRACTICING E-JUDO

1. "International Judo Federation strips titles from Vladimir Putin and Russian oligarch," World, *CBS News*, 7 March 2022, https://www.cbsnews.com/news/international-judo-federation-strips-titles-vladimir-putin-russia-ukraine-invasion/.
2. "Gallery: Lennart Meri at 90," *ERR*, 29 March 2019, https://news.err.ee/925076/gallery-lennart-meri-at-90.
3. Author's note: while in the employ of the Estonian government in 2018–2019, this was a standard talking point I was instructed to use in presentations with outside groups.
4. Lara Olszowska, "E-stonia: How the Baltic minnow became a tech powerhouse," *The Stack*, 12 September 2023, https://www.thestack.technology/estonia-technology-hub-digital/.
5. "Toomas Hendrik Ilves: Estonia as a Nordic country," delivered 14 December 1999, Swedish Institute for International Affairs, Stockholm, American Rhetoric: Online Speech Bank, https://www.americanrhetoric.com/speeches/toomasilvesestonianordiccountry.htm.
6. Anna Bocharnikova, "Economic well-being under plan versus market: The case of Estonia and Finland," *CATO Journal* (Winter 2021), https://www.cato.org/cato-journal/winter-2021/economic-well-being-under-plan-versus-market-case-estonia-finland.

NOTES

7. M. Laar, "Leading a successful transition: The Estonian miracle," *European View*, vol. 7, no. 1 (2008), 67–74, https://doi.org/10.1007/s12290-008-0024-z.
8. "Mentors, partnerships and founders with chips on their shoulders; Lux Capital's Josh Wolfe shares his perspective on all three with renegade partners," Renegade Partners, 15 March 2021, https://www.renegadepartners.com/blog/discussionwithjosh.
9. Neil Taylor, *Estonia: A Modern History* (London: C. Hurst & Co., 2018), Kindle edition, 196.
10. Toomas Hendrik Ilves, "Oral history interview," interview by Anders Hjemdahl and Camilla Andersson, Stanford Digital Repository, 1 October 2017, audio, 29:08, https://purl.stanford.edu/kz740zk2852.
11. Mart Laar, "The Estonian economic miracle," *The Heritage Foundation*, 7 August 7, 2007, https://www.heritage.org/report/the-estonian-economic-miracle.
12. Vello Ederma, "Mister Estonia," in Lennart Meri, *The Pathfinder*, Tiit Pruuli (ed.) (Tallinn, Varrak, 2009) 165.
13. Neil Taylor, *Estonia: A Modern History* (London: C. Hurst & Co., 2018), Kindle edition, 190.
14. Patrick Kingsley, "How tiny Estonia stepped out of USSR's shadow to become an internet titan," *The Guardian*, 15 April 2012, https://www.theguardian.com/technology/2012/apr/15/estonia-ussr-shadow-internet-titan.
15. Patrick Kingsley, "How tiny Estonia stepped out of USSR's shadow to become an internet titan," *The Guardian*, 15 April 2012, https://www.theguardian.com/technology/2012/apr/15/estonia-ussr-shadow-internet-titan.
16. Tarmo Kalvet, "The Estonian information society developments since the 1990s," Praxis, Working Paper No. 29, August 2007, 17, https://www.praxis.ee/wp-content/uploads/2014/03/2007-Estonian-information-society-developments.pdf.
17. Tarmo Kalvet, "The Estonian information society developments since the 1990s," Praxis, Working Paper No. 29, August 2007, 17, https://www.praxis.ee/wp-content/uploads/2014/03/2007-Estonian-information-society-developments.pdf.
18. For example, several members of the e-Residency team based in Tallinn were foreign nationals with little connection to the country and the head of the Work in Estonia program (as of writing) is an immigrant.
19. Kattel, Rainer, and Ines Mergel, "Estonia's digital transformation: Mission mystique and the hiding hand," in Paul 't Hart and Mallory Compton (eds), *Great Policy Successes* (Oxford, 2019; online edn, Oxford Academic, 24 October 2019), https://doi.org/10.1093/oso/9780198843719.003.0008.
20. "Tiger," Tiigrihupe, Education and Youth Board, https://kompass.harno.ee/tiigrihupe/algus/.
21. Sten Tamkivi, "Oral history interview with Sten Tamkivi," Baltic Video Testimonies at Stanford Libraries, 25 May 2016, video, 26:29, https://purl.stanford.edu/rf442bt3219 (lightly edited for clarity).

22. Author interview: Steve Jurvetson, email correspondence, 13 May 2024.
23. Una Bergmane, *Politics of Uncertainty: The United States, the Baltic Question, and the Collapse of the Soviet Union* (New York, NY: Oxford University Press, 2023), 172.
24. Helen Wright, "How Estonia is helping Ukraine develop e-governance," *ERR*, 9 April 2020, https://news.err.ee/1117599/feature-how-estonia-is-helping-ukraine-develop-e-governance.
25. Peeter Vihma, *Twenty Years of Building Digital Societies: Thinking about the Past and Future of Digital Transformation* (Tallinn: e-Governance Academy, 2023), 83.
26. Sten Hakewitz, "Transparency: Estonia is one of the least corrupt countries in the world," *Estonian World*, 1 February 2023, https://estonianworld.com/business/transparency-estonia-is-one-of-the-least-corrupt-countries-in-the-world/.
27. Rowland Manthorpe, "From the fires of revolution, Ukraine is reinventing government," *Wired*, 20 August 2018, https://www.wired.co.uk/article/ukraine-revolution-government-procurement.
28. Hannes Astok, "Introduction to e-government," *e-Governance Academy*, https://ega.ee/wp-content/uploads/2020/02/ega_hannese_raamat_FINAL_web.pdf.
29. Kostiantyn Gridin, Nina Kurochka, "Opinion: How Ukraine is becoming the world's number one digital government," *Kyiv Post*, 23 October 2023, https://www.kyivpost.com/opinion/23136.
30. Ott Kaukver, "Oral history interview with Ott Kaukver," *Baltic Video Testimonies at Stanford Libraries*, 26 August 2017, video, 19:04, https://purl.stanford.edu/hv021wq1614 (lightly edited for clarity).

6. THE BEDROCK OF THE E-STATE

1. Kristjan Vassil, "Estonian e-government ecosystem: Foundation, applications, outcomes," *World Bank*, June 2015, https://thedocs.worldbank.org/en/doc/165711456838073531-0050022016/original/WDR16BPEstonianeGovecosystemVassil.pdf.
2. Nathan Heller, "Estonia, The Digital Republic," *The New Yorker*, 11 December 2017, https://www.newyorker.com/magazine/2017/12/18/estonia-the-digital-republic.
3. Masha Gessen, "Why Estonia was poised to handle how a pandemic would change everything, *The New Yorker*, 24 March 2020, https://www.newyorker.com/news/our-columnists/why-estonia-was-poised-to-handle-how-a-pandemic-would-change-everything.
4. Matt Reynolds, "Welcome to E-stonia, the world's most digitally advanced society," *Wired*, 20 October 2016, https://www.wired.com/story/digital-estonia/.
5. "Principles of Estonian Information Policy," Issuu, e-Governance Academy, https://issuu.com/e-governanceacademy/docs/eesti_infopoliitika_p__hial-used.
6. Intrapreneur: someone within an organization who acts in an entrepreneurial manner, often advocating for and engaging in significant and disruptive changes.

7. Justin Petrone, "6 lessons in building a digital society," *e-Estonia*, 2 October 2023, https://e-estonia.com/lessons-and-opportunities-for-e-estonia-discussed-at-tallinn-digital-summit/.
8. Toomas Hendrik Ilves, "Oral history interview," interview by Anders Hjemdahl and Camilla Andersson, Stanford Digital Repository, 1 October 2017, audio, 1:00:42, https://purl.stanford.edu/kz740zk2852.
9. Kaja Kallas, "How Estonia Became a model for digital democracy," interview by Azeem Azhar, *Exponentially with Azeem Azhar*, *Bloomberg*, 7 September 2023, video, 13:16, https://www.bloomberg.com/news/articles/2023-09-07/video-how-estonia-became-a-digital-democracy-and-avoided-toxic-social-media.
10. For example, the e-Residency program was enabled by legislative changes that were created bespoke so that the program could get started after the concept was proposed by Ruth Annus, Taavi Kotka, and Siim Sikkut during a Development Fund competition.
11. Tarmo Kalvet and Ain Aaviksoo, "The development of eServices in an enlarged EU: eGovernment and eHealth in Estonia," *JRC Scientific and Technical Reports* (2016), 56, https://www.google.com/url?sa=t&source=web&rct=j&opi=89978449&url=https://publications.jrc.ec.europa.eu/repository/bitstream/JRC40679/jrc40679.pdf&ved=2ahUKEwjVkva0jp-FAxW1EVkFHYU0CA8QFnoECBIQAQ&usg=AOvVaw0NuLQ7cT6DisHAKOUNgk5V.
12. Justin Petrone, "6 lessons in building a digital society," *e-Estonia*, 2 October 2023, https://e-estonia.com/lessons-and-opportunities-for-e-estonia-discussed-at-tallinn-digital-summit/.
13. Justin Petrone, "6 lessons in building a digital society," *e-Estonia*, 2 October 2023, https://e-estonia.com/lessons-and-opportunities-for-e-estonia-discussed-at-tallinn-digital-summit/.
14. "DocuSign, Inc." Yahoo!finance, https://finance.yahoo.com/quote/DOCU/.
15. Kristjan Vassil, "Estonian e-Government ecosystem: Foundation, applications, outcomes," *World Bank*, June 2015, 6, https://thedocs.worldbank.org/en/doc/165711456838073531-0050022016/original/WDR16BPEstonianeGovecosystemVassil.pdf.
16. Blessing Oyetunde, "RaulWalter Estonia's digital identity giant," *e-Estonia*, 12 February 2024, https://e-estonia.com/the-making-of-a-giant-estonia-and-its-digital-identity-infrastructure/.
17. Nathan Heller, "Estonia, the digital republic," *The New Yorker*, 11 December 2017, https://www.newyorker.com/magazine/2017/12/18/estonia-the-digital-republic.
18. Patrick Kingsley, "How tiny Estonia stepped out of the USSR's shadow to become an internet titan," *The Guardian*, 15 April 2012, https://www.theguardian.com/technology/2012/apr/15/estonia-ussr-shadow-internet-titan.
19. Leonie Cater, "What Estonia's digital ID scheme can teach Europe," *Politico*, 12 March 2021, https://www.politico.eu/article/estonia-digital-id-scheme-europe/.

20. Olivia Solon, "Uganda's sweeping surveillance state is built on national ID cards," *Bloomberg*, 4 June 2024, https://www.bloomberg.com/news/features/2024-06-04/uganda-yoweri-museveni-s-critics-targeted-via-biometric-id-system.
21. Nathan Heller, "Estonia, the digital republic," *The New Yorker*, 11 December 2017, https://www.newyorker.com/magazine/2017/12/18/estonia-the-digital-republic.
22. "History of NIIS," Nordic Institute for Interoperability Solutions, https://www.niis.org/history.
23. Luukas Ilves, "The next round of European integration hinges on our ability to do GovTech together. It will not be easy," *The European Files*, 8 February 2023, https://www.europeanfiles.eu/digital/the-next-round-of-european-integration-hinges-on-our-ability-to-do-govtech-together-it-will-not-be-easy.
24. Leonie Cater, "What Estonia's digital ID scheme can teach Europe," *Politico*, 12 March 2021, https://www.politico.eu/article/estonia-digital-id-scheme-europe/.
25. Sten Tamkivi, "Lessons from the world's most tech-savvy government," *The Atlantic*, 24 January 2014, https://www.theatlantic.com/international/archive/2014/01/lessons-from-the-worlds-most-tech-savvy-government/283341/.
26. Alyssa Rosenberg, "Opinion: Parents are drowning in paperwork. That's a real problem," *The Washington Post*, 21 February 2024, https://www.washingtonpost.com/opinions/2024/02/21/parents-children-how-to-reduce-paperwork/.
27. Dr. Keegan McBride, Prof. Dr. Gerhard Hammerschmid, Hendrik Lume, and Andres Raieste, "Proactive public services—The new standard for digital governments," Hertie School Centre for Digital Governance and Nortal, May 2023, https://nortal.com/wp-content/uploads/2023/06/white-paper_proactive_public_services_en.pdf.
28. Priit Alamäe, "Challenging the 'but' in digitalization," *Nortal*, 10 January 2024, https://nortal.com/insights/challenging-the-but-in-digitalization/.
29. Taavi Rõivas: Prime Minister 2014–2016, Minister of Social Affairs 2012–2014.
30. Joel Tito, "Ever evolving, e-Estonia," Centre for Public Impact, 25 May 2017, https://www.centreforpublicimpact.org/insights/ever-evolving-e-estonia.
31. Naeha Rashid, "Deploying the once-only policy: A privacy-enhancing guide for policymakers and civil society actors," Policy Briefs Series Harvard Kennedy School Ash Center for Democratic Governance and Innovation (November 2020): 2, https://ash.harvard.edu/files/ash/files/deploying-once-only-policy.pdf?m=1605912398.
32. George Ingram and Meagan Dooley, "Digital government: Foundations for global development and democracy," Brookings Center for Sustainable Development, Working Paper #163, Global Economy and Development, December 2021.

33. Asha Barbaschow, "e-Estonia: What is all the fuss about?," *ZDNet*, 12 August 2018, https://www.zdnet.com/article/e-estonia-what-is-all-the-fuss-about/.
34. Sandra Roosna, Raul Rikk, Annela Kiirats, Anu Vahtra-Hellat, Arvo Ott, Hannes Astok, Katrin Nyman-Metcalf, Kristina Reinsalu, Liia Hänni, Mari Pedak, Olav Harjo, Raul Kaidro, Siret Schutting, and Uuno Vallner, "e-Governance in Practice e-Governance in practice," *e-Governance Academy*, https://ega.ee/wp-content/uploads/2016/06/e-Estonia-e-Governance-in-Practice.pdf.
35. Nicholas Jackson, "United Nations declares internet access a basic human right," *The Atlantic*, 3 June 2011, https://www.theatlantic.com/technology/archive/2011/06/united-nations-declares-internet-access-a-basic-human-right/239911/.
36. Allie Funk, Adrian Shahbaz, and Kian Vesteinsson, "Freedom on the net 2023: The repressive power of artificial intelligence," Freedom House, https://freedomhouse.org/report/freedom-net/2023/repressive-power-artificial-intelligence.

7. BECOMING E-ESTONIA

1. "Healthcare.gov: Ineffective planning and oversight practices underscore the need for improved contract management," U.S. Government Accountability Office, 30 July 2014, https://www.gao.gov/products/gao-14-694.
2. Elana Berkowitz and Blaise Warren, "E-government in Estonia," McKinsey, 1 June 2012, https://www.mckinsey.com/industries/public-sector/our-insights/innovation-in-government-india-and-estonia.
3. Sandra Roosna, Raul Rikk, Annela Kiirats, Anu Vahtra-Hellat, Arvo Ott, Hannes Astok, Katrin Nyman-Metcalf, Kristina Reinsalu, Liia Hänni, Mari Pedak, Olav Harjo, Raul Kaidro, Siret Schutting, and Uuno Vallner, "e-Governance in Practice," *e-Governance Academy*, https://ega.ee/wp-content/uploads/2016/06/e-Estonia-e-Governance-in-Practice.pdf.
4. Tim Maurer, "Why the Russian government turns a blind eye to cybercriminals," Slate via Carnegie Endowment for International Peace, 2 February 2018, https://carnegieendowment.org/2018/02/02/why-russian-government-turns-blind-eye-to-cybercriminals-pub-75499.
5. "World-first 'Cybercrime Index' ranks countries by cybercrime threat level," News & Events, University of Oxford, 10 April 2024, https://www.ox.ac.uk/news/2024-04-10-world-first-cybercrime-index-ranks-countries-cybercrime-threat-level.
6. Kattel, Rainer and Ines Mergel, "Estonia's digital transformation: Mission mystique and the hiding hand," in Paul 't Hart and Mallory Compton (eds), *Great Policy Successes* (Oxford, 2019; online edn, Oxford Academic, 24 October 2019), https://doi.org/10.1093/oso/9780198843719.003.0008.
7. Maximiliaan van de Poll, "The history of digital identity in Estonia," *Cybernetica*, 17 February 2020, https://cyber.ee/resources/news/the-history-of-digital-identity-in-estonia/.

8. SK ID Solutions, LinkedIn post, https://www.linkedin.com/posts/sk-id-solutions_skidsolutions-digitalidentity-smartid-activity-7046496259846680576-chyt/?originalSubdomain=ee.
9. GDP per capita, PPP (current international $)—Estonia, Finland, The World Bank, https://data.worldbank.org/indicator/NY.GDP.PCAP.PP.CD?locations=EE-FI.
10. Peter Roudik and U.S. Global Legal Research Directorate Law Library of Congress. *Estonia: Taxation System and Implementation of Flat Income Tax* (Washington, D.C.: The Law Library of Congress, Global Legal Research Directorate, 2006), www.loc.gov/item/2019669497/.
11. C. Williams, "E-tax: Electronic tax filing, Estonia," (2021), https://www.researchgate.net/publication/351984837_E-TAX_ELECTRONIC_TAX_FILING_ESTONIA.
12. Peeter Vihma, "Digitising taxation secures Estonia's #1 position in Tax Competitiveness Index," *e-Estonia*, 25 January 2023, https://e-estonia.com/digitising-taxation-secures-estonias-nr-1-position-in-tax-competitiveness-index/.
13. Justin Elliott and Paul Kiel, "Inside TurboTax's 20-year fight to stop Americans from filing their taxes for free," *ProPublica*, 17 October 2019, https://www.propublica.org/article/inside-turbotax-20-year-fight-to-stop-americans-from-filing-their-taxes-for-free.
14. "Doing Business 2020: Economy profile: Germany," World Bank Group, https://www.doingbusiness.org/content/dam/doingBusiness/country/g/germany/DEU.pdf.
15. Dylan Matthews, "I got to see the IRS's free tax-filing software in action. Here's what I learned," *Vox*, 26 February 2024, https://www.vox.com/future-perfect/24071005/irs-direct-file-free-tax-software-turbotax-review.
16. "IRS Direct File Pilot exceeds usage goal…" Press Releases, U.S. Department of the Treasury, https://home.treasury.gov/news/press-releases/jy2298?utm_source=newsletter&utm_medium=email&utm_campaign=newsletter_axiosmarkets&stream=business.
17. Sandra Roosna, Raul Rikk, Annela Kiirats, Anu Vahtra-Hellat, Arvo Ott, Hannes Astok, Katrin Nyman-Metcalf, Kristina Reinsalu, Liia Hänni, Mari Pedak, Olav Harjo, Raul Kaidro, Siret Schutting, and Uuno Vallner, "e-Governance in practice," *e-Governance Academy*, https://ega.ee/wp-content/uploads/2016/06/e-Estonia-e-Governance-in-Practice.pdf, 11.
18. "EV100: Kui e-maksuametist ei osanud veel undki näha," Ministry of Economic Affairs and Communications, 20 February 2018, https://medium.com/digiriik/meenutusi-ajaloost-kui-e-maksuametist-ei-osanud-veel-undkin%C3%A4ha-ehk-aruanne-vorm-nr-1-re-27f4790781c2.
19. "Why tax declaration is a favourite sport for Estonians? How much time would your company save on tax compliance if it paid taxes in Estonia?," *Invest in Estonia*, February 2021, https://investinestonia.com/why-tax-declaration-is-

20. Kristjan Vassil, "Estonian e-Government ecosystem: Foundation, applications, outcomes," *World Bank*, June 2015, 11–21, https://thedocs.worldbank.org/en/doc/165711456838073531-0050022016/original/WDR16BPEstonianeGovecosystemVassil.pdf.
21. Peeter Vihma, "Digitising taxation secures Estonia's #1 position in Tax Competitiveness Index," *e-Estonia*, 25 January 2023, https://e-estonia.com/digitising-taxation-secures-estonias-nr-1-position-in-tax-competitiveness-index/.
22. Emily Peck and Andrew Solender, "IRS opening free online tax filing program to all states," *Axios*, 30 May 2024, https://www.axios.com/2024/05/30/irs-taxes-direct-file-free-program.
23. Peeter Vihma, "Digitising taxation secures Estonia's #1 position in Tax Competitiveness Index," *e-Estonia*, 25 January 2023, https://e-estonia.com/digitising-taxation-secures-estonias-nr-1-position-in-tax-competitiveness-index/.
24. "Estonia first to allow online voting nationwide," *NBC News*, 14 October 2005, https://www.nbcnews.com/id/wbna9697336#.UuLMW4o5hE.
25. Paul G. Nixon and Vassiliki N. Koutrakou, *E-Government in Europe: Re-Booting the State* (New York, NY: Routledge, 2006), http://www.untag-smd.ac.id/files/Perpustakaan_Digital_1/E-GOVERNMENT%20E-government%20in%20Europe.pdf.
26. David Mac Dougall, "Estonia election: i-voting comes of age in the world's 'digital republic' with record ballots," *Euronews.next*, 3 August 2023, https://www.euronews.com/next/2023/03/08/estonia-election-i-voting-comes-of-age-in-the-worlds-digital-republic-with-record-ballots.
27. David Keyton, "High-tech Estonia votes online for European Parliament," *AP*, 20 May 2019, https://apnews.com/article/business-russia-technology-european-parliament-europe-3feb42f649de47788c5bf45a27a6cba5.
28. Kalev Aasmae, "Online voting: Now Estonia teaches the world a lesson in electronic elections," *ZDNet*, 8 March 2019, https://www.zdnet.com/article/online-voting-now-estonia-teaches-the-world-a-lesson-in-electronic-elections/.
29. "Estonia sets new e-voting record at Riigikogu 2023 elections," *ERR*, 4 March 2023, https://news.err.ee/1608904730/estonia-sets-new-e-voting-record-at-riigikogu-2023-elections.
30. Adam Rang, "Here's proof that Estonians really can vote online from anywhere," *Estonian World*, 26 February 2019, https://estonianworld.com/technology/heres-proof-that-estonians-really-can-vote-online-from-anywhere/.
31. Hannes Astok, "Introduction to e-government," *e-Governance Academy*, https://ega.ee/wp-content/uploads/2020/02/ega_hannese_raamat_FINAL_web.pdf.
32. Estonia sets new e-voting record at Riigikogu 2023 elections," *ERR*, 4 March 2023, https://news.err.ee/1608904730/estonia-sets-new-e-voting-record-at-riigikogu-2023-elections.

33. Adam Rang, "Here's proof that Estonians really can vote online from anywhere," *Estonian World*, 26 February 2019, https://estonianworld.com/technology/heres-proof-that-estonians-really-can-vote-online-from-anywhere/.
34. Miles Parks, "The push for internet voting continues, mostly thanks to one guy," *NPR*, 30 September 2021, https://www.npr.org/2021/09/30/1040999446/internet-voting-phones-tusk-grant.
35. "IVXV online voting system," Github, https://github.com/valimised/ivxv.
36. David Keyton, "High-tech Estonia votes online for European Parliament," *AP*, 20 May 2019, https://apnews.com/article/business-russia-technology-european-parliament-europe-3feb42f649de47788c5bf45a27a6cba5.
37. David Mac Dougall, "Estonia election: i-voting comes of age in the world's 'digital republic' with record ballots," *Euronews.next*, 3 August 2023, https://www.euronews.com/next/2023/03/08/estonia-election-i-voting-comes-of-age-in-the-worlds-digital-republic-with-record-ballots.
38. Estonia sets new e-voting record at Riigikogu 2023 elections," *ERR*, 4 March 2023, https://news.err.ee/1608904730/estonia-sets-new-e-voting-record-at-riigikogu-2023-elections.
39. "Statistics about Internet voting in Estonia," Valimised, https://www.valimised.ee/en/archive/statistics-about-internet-voting-estonia
40. "About," Mobile Voting, https://mobilevoting.org/about/.
41. Martin Austermuhle, "There's a new push to let D.C. voters cast ballots from their phones," *DCist*, 2 December 2021, https://dcist.com/story/21/12/02/theres-a-new-push-to-let-dc-voters-cast-ballots-from-their-phones/.
42. "Why mobile voting," Mobile Voting, https://mobilevoting.org/why-mobile-voting/.
43. Patrick Mulholland, "Estonia leads world in making digital voting a reality," *Financial Times*, 26 January 2021, https://www.ft.com/content/b4425338-6207-49a0-bbfb-6ae5460fc1c1.
44. Patrick Mulholland, "Estonia leads world in making digital voting a reality," *Financial Times*, 26 January 2021, https://www.ft.com/content/b4425338-6207-49a0-bbfb-6ae5460fc1c1.
45. Marie Jourdain, "Five takeaways from the Estonian elections, where security trumped inflation by a landslide," *Atlantic Council*, 6 March 2023, https://www.atlanticcouncil.org/blogs/new-atlanticist/five-takeaways-from-the-estonian-elections-where-security-trumped-inflation-by-a-landslide/.
46. Maarja Toots, Tarmo Kalvet, and Robert Krimmer, "Success in eVoting—Success in eDemocracy? The Estonian paradox," 8th International Conference on Electronic Participation (ePart), September 2016, Guimarães, Portugal, 55–66, https://doi.org/10.1007/978-3-319-45074-2_5. hal-01637228.
47. Piret Ehin, Mihkel Solvak, Jan Willemson, and Priit Vinkel, "Internet voting in Estonia 2005–2019: Evidence from eleven elections," *Government Information Quarterly*, vol. 39, no. 4 (2022), 11, https://www.sciencedirect.com/science/article/pii/S0740624X2200051X.
48. Piret Ehin, Mihkel Solvak, Jan Willemson, and Priit Vinkel, "Internet voting in

Estonia 2005–2019: Evidence from eleven elections," *Government Information Quarterly*, vol. 39, no. 4 (2022), 9, https://www.sciencedirect.com/science/article/pii/S0740624X2200051X.

49. Erika Piirmets, "How did Estonia carry out the world's first mostly online national elections," *e-Estonia*, 7 March 2023, https://e-estonia.com/how-did-estonia-carry-out-the-worlds-first-mostly-online-national-elections/.
50. Piret Ehin, "Reestablishing trust in democratic governments," 15 November 2024, Institute for Humane Studies at George Mason University, Lecture.
51. "Women live 11 years longer than men," Statistics Estonia, 8 March 2010, https://www.stat.ee/en/uudised/news-release-2010-036.
52. "Mortality and life expectancy statistics," *eurostat*, https://ec.europa.eu/eurostat/statistics-explained/index.php?title=Mortality_and_life_expectancy_statistics.
53. Janek Metsallik, Peeter Ross, Dirk Draheim, and Gunnar Piho, "Ten years of the e-health system in Estonia," (2018), 3, https://www.researchgate.net/publication/372890514_Ten_years_of_the_e-health_system_in_Estonia.
54. "Learning from the Estonian e-health system," *Health Europa*, 11 January 2019, https://www.healtheuropa.com/estonian-e-health-system/89750/.
55. "e-Health," Solutions, e-Estonia, https://e-estonia.com/solutions/healthcare/e-health-records/.
56. "Country vignette: Estonia: Digital technologies ensuring continuity in access to essential medicines during the Covid-19 pandemic," World Health Organization: Europe, 23 August 2021, https://www.who.int/europe/publications/m/item/estonia-digital-technologies-ensuring-continuity-in-access-to-essential-medicines-during-the-covid-19-pandemic-(2021).
57. Janek Metsallik, Peeter Ross, Dirk Draheim, and Gunnar Piho, "Ten years of the e-health system in Estonia," (2018), 2, https://www.researchgate.net/publication/372890514_Ten_years_of_the_e-health_system_in_Estonia.
58. Baoping Shang and Eva Jenkner, "The challenge of health care reform in Estonia, Hungary, China, Chile, and Mexico," International Monetary Fund, 327, https://www.elibrary.imf.org/downloadpdf/book/9781616352448/ch017.xml#:~:text=Public%20health%20spending%20as%20a%20share%20of%20GDP%20decreased%20during,2000%20to%20%24980%20in%202008.
59. Baoping Shang and Eva Jenkner, "The challenge of health care reform in Estonia, Hungary, China, Chile, and Mexico," International Monetary Fund, 329, https://www.elibrary.imf.org/downloadpdf/book/9781616352448/ch017.xml#:~:text=Public%20health%20spending%20as%20a%20share%20of%20GDP%20decreased%20during,2000%20to%20%24980%20in%202008.
60. Shio Bo and So Morikawa, "Analysis of factors influencing introduction of e-health: A case study of Estonia," *International Journal of Management and Applied Science*, vol. 3, no. 8 (August 2017), https://ijmas.iraj.in/paper_detail.php?paper_id=8918.
61. Heather Landi, "VA renegotiates $10B Oracle Cerner EHR contract with stron-

ger performance metrics, bigger penalties," *Fierce Healthcare*, 17 May 2023, https://www.fiercehealthcare.com/health-tech/va-renegotiates-10b-ehr-contract-stronger-performance-metrics-bigger-penalties.

62. Arthur Allen, "Lost in translation: Epic goes to Denmark," *Politico*, 6 June 2019, https://www.politico.com/story/2019/06/06/epic-denmark-health-1510223.
63. Olga Khazan, "What the U.S. medical system can learn from Estonia," *The Atlantic*, 25 June 2019, https://www.theatlantic.com/health/archive/2019/06/why-arent-electronic-health-records-better/592387/.
64. Olga Khazan, "What the U.S. medical system can learn from Estonia," *The Atlantic*, 25 June 2019, https://www.theatlantic.com/health/archive/2019/06/why-arent-electronic-health-records-better/592387/.
65. "Estonian e-Ambulance and time-critical health data," Cases, Scoop4C, https://scoop4c.eu/cases/estonian-e-ambulance-and-time-critical-health-data.
66. Riina Sikkut: Minister of Health and Labor 2018–2019, Minister of Economic Affairs and Infrastructure 2022–2023, Minister of Health 2023–current. Partner of Siim Sikkut, longtime public servant and CIO of Estonia 2017–2022.
67. "Learning from the Estonian e-health system," *Health Europa*, 11 January 2019, https://www.healtheuropa.com/estonian-e-health-system/89750/.
68. Sten Tamkivi, "Lessons from the world's most tech-savvy government," *The Atlantic*, 24 January 2014, https://www.theatlantic.com/international/archive/2014/01/lessons-from-the-worlds-most-tech-savvy-government/283341/.
69. Janek Metsallik, Peeter Ross, Dirk Draheim, and Gunnar Piho, "Ten years of the e-health system in Estonia," (2018), 6, https://www.researchgate.net/publication/372890514_Ten_years_of_the_e-health_system_in_Estonia.
70. "An overview of e-health services in Estonia," e-Estonia, video: 4:11, https://www.youtube.com/watch?v=H4QLzQGMI3k.
71. Janek Metsallik, Peeter Ross, Dirk Draheim, and Gunnar Piho, "Ten years of the e-health system in Estonia," (2018), 8, https://www.researchgate.net/publication/372890514_Ten_years_of_the_e-health_system_in_Estonia.
72. Janek Metsallik, Peeter Ross, Dirk Draheim, and Gunnar Piho, "Ten years of the e-health system in Estonia," (2018), 7, https://www.researchgate.net/publication/372890514_Ten_years_of_the_e-health_system_in_Estonia.
73. A. Tuula, K. Sepp, and D. Volmer, "E-solutions in Estonian community pharmacies: A literature review," *Digital Health* (2022), 8, https://doi.org/10.1177/20552076221111373122.
74. Amy Lewin, "Inside Estonia's pioneering digital health service," *Sifted*, 8 July 2020, https://sifted.eu/articles/estonia-digital-health.
75. "Estonia and WHO to work together on digital health and innovation," News, World Health Organization: Europe, 7 October 2020, https://www.who.int/europe/news/item/07–10–2020-estonia-and-who-to-work-together-on-digital-health-and-innovation.

76. "Promoting the development of the Moldovan health care system with the opportunities of e-services," Praxis, https://www.praxis.ee/en/tood/promoting-the-development-of-the-moldovan-health-care-system-with-the-opportunities-of-e-services/.
77. Dr. Thomas Kostera, "#SmartHealthSystems: International comparison of digital strategies," *Bertelsmann Stiftung*, https://www.bertelsmann-stiftung.de/fileadmin/files/Projekte/Der_digitale_Patient/VV_SHS_Europe_eng.pdf.
78. Masha Gessen, "Why Estonia was poised to handle how a pandemic would change everything," *The New Yorker*, 24 March 2020, https://www.newyorker.com/news/our-columnists/why-estonia-was-poised-to-handle-how-a-pandemic-would-change-everything/.
79. "Estonia: Country health profile," OECD, 28 November 2019, https://www.oecd.org/publications/estonia-country-health-profile-2019-0b94102e-en.htm.
80. "State of health in the EU: Estonia: Country health profile 2021," OECD, 13 December 2021, https://www.oecd.org/estonia/estonia-country-health-profile-2021-a6c1caa5-en.htm.
81. "Mortality and life expectancy statistics," Eurostat: Statistics Explained, https://ec.europa.eu/eurostat/statistics-explained/index.php?title=Mortality_and_life_expectancy_statistics.
82. Silviu Kondan, Mridvika Sahajpal, and David J. Trimbach, "Identifying the needs of Estonia's Russian-speaking minority: COVID-19, data disaggregation, and social determinants of health," Foreign Policy Research Institute, 11 May 2021, https://www.fpri.org/article/2021/05/identifying-the-needs-of-estonias-russian-speaking-minority-covid-19-data-disaggregation-and-social-determinants-of-health/#:~:text=Compared%20to%20ethnic%20Estonians%2C%20Russian,to%20lower%20Estonian%20language%20proficiencies.
83. Silviu Kondan, Mridvika Sahajpal, and David J. Trimbach, "Identifying the needs of Estonia's Russian-speaking minority: COVID-19, data disaggregation, and social determinants of health," Foreign Policy Research Institute, 11 May 2021, https://www.fpri.org/article/2021/05/identifying-the-needs-of-estonias-russian-speaking-minority-covid-19-data-disaggregation-and-social-determinants-of-health/#:~:text=Compared%20to%20ethnic%20Estonians%2C%20Russian,to%20lower%20Estonian%20language%20proficiencies.
84. "State of health in the EU: Estonia: Country health profile 2021," OECD, 13 December 2021, https://www.oecd.org/estonia/estonia-country-health-profile-2021-a6c1caa5-en.htm.
85. Silver Tambur, "Dr Keegan McBride: The current government is making the same mistakes the previous one did," *Estonian World*, 8 March 2021, https://estonianworld.com/security/dr-keegan-mcbride-the-current-government-is-making-the-same-mistakes-the-previous-one-did/.
86. Dr. Keegan McBride, "Opinion: Pandemic highlighted shortcomings in e-Government," *ERR*, 1 July 2021, https://news.err.ee/1229992/opinion-pandemic-highlighted-shortcomings-in-estonian-e-government.

87. "Precious failures," *e-Estonia*, 18 August 2021, https://e-estonia.com/precious-failures/.
88. "Remarks by President Obama and President Ilves of Estonia in joint press conference," Office of the Press Secretary, White House, 3 September 2014, https://obamawhitehouse.archives.gov/the-press-office/2014/09/03/remarks-president-obama-and-president-ilves-estonia-joint-press-confer-0.
89. Ben Horowitz and Sten Tamkivi, "Estonia: The little country that cloud," Andreessen Horowitz Blog, 6 February 2014, https://a16z.com/estonia-the-little-country-that-cloud/
90. Giulia Lanzuolo, "Alluring narratives: What do Estonia, Bhutan and Singapore have in common?," UCL Institute for Innovation and Public Purpose IIPP Student Ideas, 4 April 2023, https://medium.com/iipp-mpa-blog/alluring-narratives-what-do-estonia-bhutan-and-singapore-have-in-common-873ce92e1ee1.

8. PLAYING LEAPTIGER

1. Steven Vaughan-Nichols, "Mosaic turns 25: The beginning of the modern web," *ZDNet*, 25 April 2018, https://www.zdnet.com/home-and-office/networking/mosaics-birthday-25-years-of-the-modern-web/.
2. "About," About, Andreessen Horowitz, https://a16z.com/about/.
3. Ben Gilbert and David Rosenthal, "Skype," Season 1, Episode 24, https://www.acquired.fm/episodes/episode-24-skype.
4. Toomas Hendrik Ilves, "Oral history interview with Toomas Hendrick Ilves," *Baltic Video Testimonies at Stanford Libraries*, 1 October 2017, video, 40:40, https://purl.stanford.edu/kz740zk2852.
5. Patrick Kingsley, "How tiny Estonia stepped out of USSR's shadow to become an internet titan," *The Guardian*, 15 April 2012, https://www.theguardian.com/technology/2012/apr/15/estonia-ussr-shadow-internet-titan.
6. Henrik Roonemaa, "Tiigrihüpe," Republic of Estonia Education and Youth Board, https://kompass.harno.ee/tiigrihupe/algus/.
7. Henry Kissenger, "Foreword," *From Third World to First: The Singapore Story: 1965–2000* (New York, NY: HarperCollins Publishers, 2000), X.
8. Henrik Roonemaa, "Tiigrihüpe," Republic of Estonia Education and Youth Board, https://kompass.harno.ee/tiigrihupe/algus/.
9. Toomas Hendrik Ilves. "Former Estonian President Toomas Hendrik Ilves Q&A," *Finding ctrl*, Nesta, https://findingctrl.nesta.org.uk/toomas-hendrik-ilves/.
10. "Most 'startup-friendly' and unicorns galore: The Baltics in data," *Sifted*, 9 February 2022, https://sifted.eu/articles/baltics-estonia-latvia-lithuania-startups-ecosystem.
11. "Internet use is increasing in Estonia," News, Statistics Estonia, 14 September 2023, https://www.stat.ee/en/news/information-technology-households-

2023#:~:text=According%20to%20Statistics%20Estonia*%2C%20 93.2,an%20internet%20connection%20at%20home.
12. Jaak Aaviksoo: Minister of Culture and Education 1995–1996, Minister of Defense 2007–2011, Minister of Education and Research 2011–2014.
13. Neil Taylor, *Estonia: A Modern History* (London: C. Hurst & Co., 2018), Kindle edition, 196.
14. Henrik Roonemaa, "Tiigrihüpe," Republic of Estonia Education and Youth Board, https://kompass.harno.ee/tiigrihupe/.
15. Tiit Pruuli, "Meaningful symbols," Lennart Meri, *The Pathfinder*, Tiit Pruuli (ed.) (Tallinn: Varrak 2009), 165.
16. Evelin Andrespok, "Estonian e-tiger leaping to Georgia: Added value of Estonian development cooperation," University of Helsinki Faculty of Social Sciences, Master's Thesis, May 2014, 10, https://helda.helsinki.fi/server/api/core/bitstreams/dda0c81f-02a2-4998-9404-ca0a862231f0/content.
17. "PISA," Statistics and Analysis, Ministry of Education and Research, https://www.hm.ee/en/ministry/statistics-and-analysis/pisa#:~:text=Estonia%20 ranks%201st%20among%20European,1st%2D2nd%20position%20with%20 Ireland.
18. Henrik Roonemaa, "Tiigrihüpe," Republic of Estonia Education and Youth Board, https://kompass.harno.ee/tiigrihupe/.
19. Jaak Aaviksoo, "Interview with Jaak Aaviksoo," The History of E-Estonia: Oral History Interviews with Estonian Experts on the Development of Estonia's Digital Society 1991–2021, 14 October 2021, translated transcripts, https://searchworks.stanford.edu/view/rp867fy3044.
20. Virtual conversation with Ants Sild on 29 May 2024.
21. Sandra Roosna, Raul Rikk, Annela Kiirats, Anu Vahtra-Hellat, Arvo Ott, Hannes Astok, Katrin Nyman-Metcalf, Kristina Reinsalu, Liia Hänni, Mari Pedak, Olav Harjo, Raul Kaidro, Siret Schutting, and Uuno Vallner, "e-Governance in practice," *e-Governance Academy*, 10, https://ega.ee/wp-content/uploads/2016/06/e-Estonia-e-Governance-in-Practice.pdf.
22. Alar Ehandi, "The 'Look@World' project: An initiative from Estonia's private sector to boost internet use," *Baltic IT&T Review*, http://www.ebaltics.lv/doc_upl/Ehandi(2).pdf.
23. Pille Runnel, Pille Pruulmann-Vengerfeldt, and Kristina Reinsalu, "The Estonian tiger leap from post-communism to the information society: From policy to practice," *Journal of Baltic Studies*, vol. 40, no. 1 (2009), 29–51, https://doi.org/10.1080/01629770902722245.
24. Henrik Roonemaa, "Tiigrihüpe," Republic of Estonia Education and Youth Board, https://kompass.harno.ee/tiigrihupe/.
25. Tarmo Kalvet, "The Estonian information society developments since the 1990s," Praxis, Working Paper No. 29, August 2007, 27, https://www.praxis.ee/wp-content/uploads/2014/03/2007-Estonian-information-society-developments.pdf.

26. Henrik Roonemaa, "Tiigrihüpe," Republic of Estonia Education and Youth Board, https://kompass.harno.ee/tiigrihupe/.
27. Toomas Hendrik Ilves, "Never provincial, always connected: An interview with former president Toomas Hendrik Ilves," *Enterprise Estonia*, 17 February 2021, https://e-estonia.com/never-provincial-always-connected-an-interview-with-former-president-toomas-hendrik-ilves/.
28. "Tiger Leap: 1997–2007," *Tiger Leap Foundation*, 2007, https://www.educationestonia.org/wp-content/uploads/2023/01/tiigrihype2007ENG_standard.pdf.
29. Alar Ehandi, "The 'Look@World' project: An initiative from Estonia's private sector to boost internet use," *Baltic IT&T Review*, http://www.ebaltics.lv/doc_upl/Laanpere.pdf.
30. "How it all began? From Tiger Leap to digital society," Education Estonia, https://www.educationestonia.org/tiger-leap/.
31. Henrik Roonemaa, "Tiigrihüpe," Republic of Estonia Education and Youth Board, https://kompass.harno.ee/tiigrihupe/tiigrihupe-pluss-ehk-opetajalt-opetajale/.
32. Henrik Roonemaa, "Tiigrihüpe," Republic of Estonia Education and Youth Board, https://kompass.harno.ee/tiigrihupe/tiigrihupe-pluss-ehk-opetajalt-opetajale/.
33. Danielle Hinton, Adi Kumar, Sara O'Rourke, Kunal Modi, Blair Levin, Anne Neville-Bonilla, Larry Strickling, and Jon Wilkins, "Are states ready to close the US digital divide?," McKinsey & Company Public Sector, 1 June 2022, https://www.mckinsey.com/industries/public-sector/our-insights/are-states-ready-to-close-the-us-digital-divide.
34. Heli Aru-Chabilan, "Tiger Leap for digital turn in the Estonian education," *Educational Media International*, vol. 57, no. 1 (2020), 61–72, https://doi.org/10.1080/09523987.2020.1744858.
35. Ott Kaukver, "Oral history interview with Ott Kaukver," *Baltic Video Testimonies at Stanford Libraries*, 26 August 2017, video, 24:40, https://purl.stanford.edu/hv021wq1614.

9. THE ESTONIAN TECH MAFIA

1. "Lift99 Tallinn Hub," Matterport, https://my.matterport.com/show/?m=FMLvZLPS63Q&hl=1&tourcta=1&ts=2&play=1.
2. In-person conversation with Sten Tamkivi on 9 April 2024.
3. Craig Turp-Balazs, "Why Skype remains key to Estonia's digital success," *Emerging Europe*, 15 April 2020, https://emerging-europe.com/business/why-skype-remains-key-to-estonias-digital-success/.
4. Sten Tamkivi (@seikatsu), "9/For this weekend @atomico Insights team kindly scanned their stateofeuropeantech.com dataset for more numbers and it is amazing. Skypers have now created 225 tech companies. And if you cound the

second generation (e.g. Skype -> Wise -> Lightyear) we are at 910 companies born," Twitter, 18 September 2023, https://twitter.com/seikatsu/status/1703676031368843502

5. Steve O'Hear, "Wise's Taavet Hinrikus and Teleport's Sten Tamkivi partner in new investment firm—Just don't call it a VC fund," *TechCrunch*, 11 March 2021, https://techcrunch.com/2021/03/11/fund-with-no-name/.
6. "Taavet+Sten," Seikatsu: Sten Tamkivi's blog archive, 10 March 2021, https://sten.tamkivi.com/2021/03/taavet-sten/.
7. "Topia and Teleport," Topia, https://teleport.org/blog/2017/04/teleport-a-move-guides-company/.
8. "The Estonian taxibooking app Taxify raises $100K," *Estonian World*, 28 April 2014, https://estonianworld.com/technology/estonian-taxi%C2%ADbooking-app-taxify-raises-100k/.
9. Ryan Browne, "European Uber rival Bolt valued at $8.4 billion in new funding round," *CNBC*, 11 January 2022, https://www.cnbc.com/2022/01/11/european-uber-rival-bolt-valued-at-8point4-billion-in-new-funding-round.html#:~:text=Estonia%2Dbased%20ride%2Dhailing%20start,billion%20just%20five%20months%20ago.
10. Vicky McKeever, "How a college dropout became Europe's youngest founder of a billion-dollar company," *CNBC*, 21 October 2019, https://www.cnbc.com/2019/10/21/how-bolt-ceo-markus-villig-became-europes-youngest-unicorn-founder.html.
11. Sten Tamkivi, "Technology's impact on GDP could be outsized," *Politico*, 25 October 2022, https://www.politico.eu/article/technologys-impact-on-gdp-could-be-outsized/.
12. "State of European tech," Atomico, https://www.stateofeuropeantech.com/.
13. Amy Lewin, "Inside Taavet+Sten: Not your typical startup investor," *Sifted*, 28 April 2022, https://sifted.eu/articles/taavet-sten-startup-fund.
14. "History of the Estonian digital society," Cybernetica, 25 May 2022, https://cyber.ee/resources/stories/history-of-digital-society/.
15. Arnaud Castaignet, "Skeleton raises €108M from top investors including Siemens & Marubeni," *Skeleton Blog*, 13 October 2024, https://www.skeletontech.com/news/skeleton-technologies-secures-108m-eur-of-financing-with-top-investors-including-siemens-and-marubeni.
16. Ingrid Lunden, "Starship Technologies raises another $42M to fuel the growth of its fleet of self-driving delivery robots," *Techcrunch*, 1 March 2022, https://techcrunch.com/2022/03/01/starship-technologies-raises-another-42m-to-fuel-the-growth-of-its-fleet-of-self-driving-delivery-robots/.
17. Sten Tamkivi, "Oral history interview with Sten Tamkivi," *Baltic Video Testimonies at Stanford Libraries*, 25 May 2016, video, 30:22, https://purl.stanford.edu/rf442bt3219. Lightly edited for clarity.
18. Steve O'Hear, "Sales CRM Pipedrive takes majority investment from Vista Equity Partners to reach unicorn status," *TechCrunch*, 12 November 2020,

https://techcrunch.com/2020/11/12/european-unicorns-are-no-longer-a-pipe-dream/.
19. Agur Jõgi, "Entrepreneurial lessons from the Estonian unicorn factory," *Forbes*, 6 May 2024, https://www.forbes.com/sites/forbestechcouncil/2024/05/06/entrepreneurial-lessons-from-the-estonian-unicorn-factory/?utm_content=292090422&utm_medium=social&utm_source=linkedin&hss_channel=lcp-11075183&sh=201ac19a5c70.
20. Tiit Riisalo: Ambassador-at-Large for Connectivity January 2022–January 2023, Minister of Economic Affairs and Information Technology April 2023–present as of publication.
21. Lara Olszowska, "E-stonia: How the Baltic minnow became a tech powerhouse," *The Stack*, 12 September 2023, https://www.thestack.technology/estonia-technology-hub-digital/.
22. Rainer Kattel, "Is Estonia the Silicon Valley of digital government," UCL Institute for Innovation and Public Purpose IIPP Blog, 28 September 2018, https://medium.com/iipp-blog/is-estonia-the-silicon-valley-of-digital-government-bf15adc8e1ea.
23. Gordon Kelly, "Finland and Nokia: An affair to remember," *Wired*, 4 October 2013, https://www.wired.com/story/finland-and-nokia/.
24. Meelis Kitsing, "Explaining the e-government success in Estonia," 429–430 (January 2008), https://www.researchgate.net/publication/221584877_Explaining_the_e_government_success_in_Estonia.
25. "Taxation of share options in Estonia," Hedman Partners & Co, 25 May 2021, https://hedman.legal/articles/taxation-of-share-options/.
26. "Not Optional campaign drives €5 billion more into the hands of European startup employees," Not Optional, 2 July 2023, https://www.notoptional.eu/latest-news/not-optional-campaign-drives-eu5-billion-more-into-the-hands-of-european-startup-employees.html.
27. Kristiina Kriisa, "Milrem Robotics: Leading the way in autonomous ground systems innovation," *e-Estonia*, 18 September 2024, https://e-estonia.com/milrem-robotics-autonomous-ground-systems-innovation/.
28. Leigh Alexander and Matt Shore, "Internet access is now a basic human right: Part 2—Chips with Everything tech podcast," Chips with Everything, *The Guardian*, 4 August 2016, audio, 13:58, https://www.theguardian.com/technology/audio/2016/aug/04/internet-access-human-right-2-tech-podcast.
29. Sten-Kristian Saluveer and Maarika Truu, "Startup Estonia White Paper: 2021–2027," Startup Estonia, July 2020, https://media.voog.com/0000/0037/5345/files/SE_Whitepaper_Web%20(1)-1.pdf, 2.
30. "Support for entrepreneurs: Startup Estonia published new documents necessary for new businesses," Startup Estonia, 6 June 2023, https://startupestonia.ee/support-for-entrepreneurs-startup-estonia-published-new-documents-necessary-for-new-businesses/.
31. Sten-Kristian Saluveer and Maarika Truu, "Startup Estonia White Paper: 2021–

2027," Startup Estonia, July 2020, https://media.voog.com/0000/0037/5345/files/SE_Whitepaper_Web%20(1)-1.pdf, 12.
32. Sten-Kristian Saluveer and Maarika Truu, "Startup Estonia White Paper: 2021–2027," Startup Estonia, July 2020, https://media.voog.com/0000/0037/5345/files/SE_Whitepaper_Web%20(1)-1.pdf, 12.
33. Sten-Kristian Saluveer and Maarika Truu, "Startup Estonia White Paper: 2021–2027," Startup Estonia, July 2020, https://media.voog.com/0000/0037/5345/files/SE_Whitepaper_Web%20(1)-1.pdf, 12.
34. Charlie Duxbury, "Estonia's techies fear the far right," *Politico*, 2 December 2020, https://www.politico.eu/article/estonia-tech-companies-digital-nomads-fear-the-far-right/.
35. Charlie Duxbury, "Estonia's techies fear the far right," *Politico*, 2 December 2020, https://www.politico.eu/article/estonia-tech-companies-digital-nomads-fear-the-far-right/.
36. Andres Burgos, "Estonia legalizes same-sex marriage," *Human Rights Watch*, 22 June 2023, https://www.hrw.org/news/2023/06/22/estonia-legalizes-same-sex-marriage.
37. Elle Hunt, "Estonian President delights in country's high proportion of unicorns," *The Guardian*, 29 June 2018, https://www.theguardian.com/world/2018/jun/29/estonia-unicorns-president-kersti-kaljulaid-delight.
38. Markus Villig, "Weak national defense scaring away investors in Estonia," *ERR*, 21 December 2023, https://news.err.ee/1609201822/markus-villig-weak-national-defense-scaring-away-investors-in-estonia.
39. "NATO DIANA invests almost €1 milliion in Estonia-based startups," News, *ERR*, 12 January 2023, https://news.err.ee/1609182358/nato-diana-invests-almost-1-million-in-estonia-based-startups.
40. "Estonia creating €50 million defense investment fund," News, *ERR*, 5 February 2024, https://news.err.ee/1609330737/estonia-creating-50-million-defense-investment-fund.
41. Marek Grzegorczyk, "Why Estonian parents are winning at entrepreneurship," *Emerging Europe*, 18 November 2022, https://emerging-europe.com/news/why-estonian-parents-are-winning-at-entrepreneurship//
42. Matt Ross, "From mini-state to digital giant: Siim Sikkut on Estonia's remarkable journey," *Global Government Forum*, 2 February 20222, https://www.globalgovernmentforum.com/from-mini-state-to-digital-giant-siim-sikkut-on-estonias-remarkable-journey/.
43. Marian Männi, "University of Tartu study explains success of Estonian entrepreneurs," *ERR*, 26 April 2022, https://news.err.ee/1608577936/university-of-tartu-study-explains-success-of-estonian-entrepreneurs.
44. "Estonian startups' 2022 turnover up 49 percent on year, exceeds €2 billion," *ERR*, 2 February 2023, https://news.err.ee/1608895289/estonian-startups-2022-turnover-up-49-percent-on-year-exceeds-2-billion.
45. Kirstie Pickering, "The digital nomad's guide to Estonia," *Sifted*, 29 June 2023, https://sifted.eu/articles/digital-nomads-guide-estonia-brnd.

46. "Inclusive entrepreneurship policy country assessment notes: Estonia 2022–23," OECD, https://www.oecd.org/cfe/smes/Estonia.pdf, 7.
47. "Emerge Estonia," Our Programs / Early Stage, Tenity, https://www.tenity.com/programs/emerge-estonia.
48. Neil Taylor, *Estonia: A Modern History* (London: C. Hurst & Co., 2018), Kindle edition, 191–192.
49. Neil Taylor, *Estonia: A Modern History* (London: C. Hurst & Co., 2018), Kindle edition, 191–192.

10. EXPORTING ESTONIA

1. Silver Tambur, "Estonian software company Cybernetica modernises Palestine's digital society," *Estonian World*, 14 April 2021, https://estonianworld.com/technology/estonian-software-company-cybernetica-modernises-palestines-digital-society/.
2. Kitsing, Meelis (University of Massachusetts Amherst) "An evaluation of e-government in Estonia," prepared for delivery at the Internet, Politics and Policy 2010: An Impact Assessment conference at Oxford University, 2010, https://blogs.oii.ox.ac.uk/ipp-conference/sites/ipp/files/documents/IPP2010_Kitsing_1_Paper_0.pdf.
3. "Our story," Company, Cybernetica, https://cyber.ee/company/our-story/.
4. "Our story," Company, Cybernetica, https://cyber.ee/company/our-story/.
5. "e-Estonia podcast: Kevin Tammearu of Cybernetica: E-solutions act as tools for democracy," *e-Estonia*, https://e-estonia.com/e-estonia-podcast-cybernetica/.
6. In-person conversation with Arne Ansper on 8 April 2024.
7. Tarmo Kalvet, "The Estonian information society developments since the 1990s," Praxis, Working Paper No. 29, August 2007, 21, https://www.praxis.ee/wp-content/uploads/2014/03/2007-Estonian-information-society-developments.pdf.
8. Kattel, Rainer and Ines Mergel, "Estonia's digital transformation: Mission mystique and the hiding hand," in Paul 't Hart and Mallory Compton (eds), *Great Policy Successes* (Oxford, 2019; online edn, Oxford Academic, 24 October 2019), https://doi.org/10.1093/oso/9780198843719.003.0008.
9. "e-Estonia podcast: Kevin Tammearu of Cybernetica: E-solutions act as tools for democracy," *e-Estonia*, https://e-estonia.com/e-estonia-podcast-cybernetica/.
10. Ann-Marii Nergi, "World-class IT R&D in Estonia: Cybernetica creates IT-solutions that just cannot fail," *Invest in Estonia*, November 2020, https://investinestonia.com/world-class-it-rd-in-estonia-cybernetica-creates-it-solutions-that-just-cannot-fail/.
11. "Remote tower air traffic services opens many new opportunities for Estonian air traffic controllers both at home and abroad," News, *EANS*, 12 May 2022, https://www.eans.ee/en/uudised/irdtorni-lahendus-avab-eesti-lennujuhtidele-palju-uusi-voimalusi-nii-kodus-kui-voorsil.

12. "Adacel enters new markets with the virtual ATC Tower acquisition," *Canso*, https://canso.org/adacel-enters-new-markets-with-the-virtual-atc-tower-acquisition/.
13. Amanda Zink, "Pulling ahead of the pack: How public-private partnerships have advanced Estonia's democracy," *International Republican Institute*, 27 January 2020, https://www.iri.org/news/pulling-ahead-of-the-pack-how-public-private-partnerships-have-advanced-estonias-democracy/.
14. Patrick Kingsley, "How tiny Estonia stepped out of USSR's shadow to become an internet titan," *The Guardian*, 15 April 2012, https://www.theguardian.com/technology/2012/apr/15/estonia-ussr-shadow-internet-titan.
15. Nurul Ardhaninggar, "E-government success stories: Learning from Denmark and Estonia," *ModernDiplomacy*, 5 December 2023, https://moderndiplomacy.eu/2023/12/05/e-government-success-stories-learning-from-denmark-and-estonia/.
16. Christopher Schepers, "Maryland airmen, Estonia build cyber-sharing platform," Air National Guard, 21 June 2023, https://www.ang.af.mil/Media/Article-Display/Article/3434815/maryland-airmen-estonia-build-cyber-sharing-platform/.
17. Sam Trendall, "Estonia: How the X-Road paved the way the way to a digital society," *Public Technology*, 20 December 2023, https://www.publictechnology.net/2023/12/20/society-and-welfare/estonia-how-the-x-road-paved-the-way-to-a-digital-society/.
18. President Toomas Hendrik Ilves, "Toomas Ilves: Lessons in digital democracy from Estonia," Interview by Russ Altman, The Future of Everything, Stanford University School of Engineering, 4 February 2019, video, 28:27, https://www.youtube.com/watch?v=ISchpVmmQiU.
19. Kevin Tammearu, "What the United States can learn from Estonia on e-governance," *CEPA*, 31 August 2021, https://cepa.org/comprehensive-reports/what-the-united-states-can-learn-from-estonia-on-e-governance/.
20. Ines Mergel, "Digital service teams in government," *Government Information Quarterly*, vol. 36, no. 4 (2019), https://doi.org/10.1016/j.giq.2019.07.001.
21. "Estonian representations around the world," Consular Visa and Travel Information, Republic of Estonia Ministry of Foreign Affairs, https://www.vm.ee/en/consular-visa-and-travel-information/estonian-representations-around-world.
22. In-person conversation with Arne Ansper, 8 April 2024.
23. Elisabeth Gosselin-Malo, "Russia ups cash reward for capturing Estonian ground robot in Ukraine," *DefenseNews*, 16 February 2024, https://www.defensenews.com/global/europe/2024/02/16/russia-ups-cash-reward-for-capturing-estonian-ground-robot-in-ukraine/.
24. Eric Lipton, "New spin on a revolving door: Pentagon officials turned venture capitalists," *New York Times*, 30 December 2023, https://www.nytimes.com/2023/12/30/us/politics/pentagon-venture-capitalists.html.
25. Austin Wright, "DOD's revolving door in full swing," *Politico*, 24 October 2013,

https://www.politico.com/story/2013/10/department-of-defenses-revolving-door-in-full-swing-098813.

26. William D. Hartung, "March of the four-stars: The role of retired generals and admirals in the arms industry," *Quincy Institute for Responsible Statecraft*, 4 October 2023, https://quincyinst.org/report/march-of-the-four-stars-the-role-of-retired-generals-and-admirals-in-the-arms-industry/.
27. "Estonia to acknowledged 99 people with national decorations on the eve of Independence Day," Press Releases, Office of the President, https://vp2006–2016.president.ee/en/media/press-releases/11991-estonia-to-acknowledged-99-people-with-national-decorations-on-the-eve-of-independence-day/index.html.
28. Tanel Saarmann, "Nortal—The company that built a third of e-Estonia," *e-Estonia*, 8 October 2020, https://e-estonia.com/nortal-the-company-that-built-a-third-of-e-estonia/.
29. Justin Petrone, "Digital Discussion spotlight: Oliver Väärtnõu, CEO of Cybernetica," *e-Estonia*, 8 March 2022, https://e-estonia.com/digital-discussion-spotlight-oliver-vaartnou-ceo-of-cybernetica/.
30. "Sten Tamkivi, Head of Skype Estonia, to advise the President," Press Releases, Office of the President, 10 January 2009, https://vp2006–2016.president.ee/en/media/press-releases/1999-sten-tamkivi-head-of-skype-estonia-to-advise-the-president/index.html.
31. Jon Henley, "Estonian government collapses over corruption investigation," *The Guardian*, 13 January 2021, https://www.theguardian.com/world/2021/jan/13/estonian-government-collapses-over-corruption-investigation.
32. Laura Hülsemann, "Kaja Kallas faces more heat amid probe of huband's Russia business ties," *Politico*, 28 August 2023, https://www.politico.eu/article/kaja-kallas-husband-russian-business-ties-estonia/.
33. Siim Sikkut, *Digital Government Excellence: Lessons from Effective Digital Leaders* (Hoboken: John Wiley & Sons, 2022), 235.
34. Siim Sikkut, *Digital Government Excellence: Lessons from Effective Digital Leaders* (Hoboken: John Wiley & Sons, 2022), 243.
35. Jennifer Pahlka, "Testimony: Jennifer Pahlka on harnessing AI to improve government services and customer experience," Niskanen Center, 10 January 2024, https://www.niskanencenter.org/testimony-jennifer-pahlka-on-harnessing-ai-to-improve-government-services-and-customer-experience/.
36. Mallory Moench, "S.F. takes 255 days to hire a city worker," *San Francisco Chronicle*, 18 October 2022, https://www.sfchronicle.com/bayarea/article/s-f-takes-255-days-to-hire-a-city-worker-17515440.php.
37. Nathan Heller, "Estonia, the digital republic," *The New Yorker*, 11 December 2017, https://www.newyorker.com/magazine/2017/12/18/estonia-the-digital-republic.
38. "Public procurement—Study on administrative capacity in the EU," Estonia Country Profile, European Commission, https://ec.europa.eu/regional_pol-

icy/sources/policy/how/improving-investment/public-procurement/study/country_profile/ee.pdf.
39. "Corruption Perceptions Index: 2022," Transparency International, https://www.transparency.org/en/cpi/2022.
40. Piia Tammpuu and Anu Masso, "'Welcome to the virtual state': Estonian e-residency and the digitalised state as a commodity," *European Journal of Cultural Studies*, vol. 21 (2018), vol. 21.
41. Toomas Hendrik Ilves, "Oral history interview with Toomas Hendrick Ilves," *Baltic Video Testimonies at Stanford Libraries*, 1 October 2017, video, 19:01, https://purl.stanford.edu/kz740zk2852.
42. Rainer Kattel, Ines Mergel, "Estonia's digital transformation: Mission mystique and the hiding hand," UCL Institute for Innovation and Public Purpose Working Paper (September 2018): 10, https://www.ucl.ac.uk/bartlett/public-purpose/sites/public-purpose/files/iipp-wp-2018-09_estonias_digital_transformation.pdf.

11. WEB WAR ONE

1. Damien McGuinness, "How a cyber attack transformed Estonia," *BBC*, 27 April 2017, https://www.bbc.com/news/39655415.
2. Pascal Davies, "Estonia hit by 'most extensive' cyberattack since 2007 amid tensions with Russia over Ukraine war," *Euronews*, 18 August 2022, https://www.euronews.com/next/2022/08/18/estonia-hit-by-most-extensive-cyberattack-since-2007-amid-tensions-with-russia-over-ukrain.
3. "Estonian denial of service incident," Cyber Operations Home, Council on Foreign Relations, May 2007, https://www.cfr.org/cyber-operations/estonian-denial-service-incident.
4. Ian Traynor, "Russia accused of unleashing cyberwar to disable Estonia," *The Guardian*, 16 May 2007, https://www.theguardian.com/world/2007/may/17/topstories3.russia.
5. Ivo Juurvee and Mariita Maatiisen, "The Bronze Soldier crisis of 2007: Revisiting an early case of hybrid conflict," International Centre for Defence and Security, August 2020, https://icds.ee/wp-content/uploads/2020/08/ICDS_Report_The_Bronze_Soldier_Crises_of_2007_Juurvee_Mattiisen_August_2020.pdf.
6. Andreas Schmidt, "The Estonian cyberattacks," in Jason Healey (ed.), *The Fierce Domain—Conflicts in Cyberspace 1986–2012* (Washington, D.C.: Atlantic Council, 2013), https://www.researchgate.net/publication/264418820_The_Estonian_Cyberattacks._.
7. "Annual Review 2023–2024," Estonian Internal Security Service, https://kapo.ee/sites/default/files/content_page_attachments/Annual%20review%202023-2024.pdf.
8. Jonas Heering and Heera Kamboj, "Case 355—Estonia: The first battle in the modern disinformation war—Lessons for democracies fighting hybrid warfare," Institute for the Study of Diplomacy at Georgetown University.

9. Ivo Juurvee and Mariita Maatiisen, "The Bronze Soldier crisis of 2007: Revisiting an early case of hybrid conflict," International Centre for Defence and Security, August 2020, https://icds.ee/wp-content/uploads/2020/08/ICDS_Report_The_Bronze_Soldier_Crises_of_2007_Juurvee_Mattiisen_August_2020.pdf.
10. Andreas Schmidt, "The Estonian Cyberattacks," in Jason Healey (ed.), *The Fierce Domain—Conflicts in Cyberspace 1986–2012* (Washington, D.C.: Atlantic Council, 2013), https://www.researchgate.net/publication/264418820_The_Estonian_Cyberattacks.
11. Hannes Grassegger and Mikael Krogerus, "Fake news and botnets: How Russia weaponised the web," *The Guardian*, 2 December 2017, https://www.theguardian.com/technology/2017/dec/02/fake-news-botnets-how-russia-weaponised-the-web-cyber-attack-estonia.
12. Tom Sear, "Cyber attacks ten years on: From disruption to disinformation," *The Conversation*, 26 April 2017, https://theconversation.com/cyber-attacks-ten-years-on-from-disruption-to-disinformation-75773.
13. "NATO—Six Colours: War in cyberspace [2009]," YouTube, video, 1:55, https://www.youtube.com/watch?v=oGZkCdpPLBE&t=4s.
14. Hannes Grassegger and Mikael Krogerus, "Fake news and botnets: How Russia weaponised the web," *The Guardian*, 2 December 2017, https://www.theguardian.com/technology/2017/dec/02/fake-news-botnets-how-russia-weaponised-the-web-cyber-attack-estonia.
15. Joshua Davis, "Hackers take down the most wired country in Europe," *Wired*, 21 August 2007, https://www.wired.com/2007/08/ff-estonia/.
16. "Governance—Estonia," National Interoperability Framework Observatory, https://joinup.ec.europa.eu/collection/nifo-national-interoperability-framework-observatory/governance-estonia.
17. "NATO—Six colours: War in cyberspace [2009]," YouTube, video, 3:09, https://www.youtube.com/watch?v=oGZkCdpPLBE&t=4s.
18. Valentinas Mite, "Estonia: Attacks seen as first case of 'cyberwar,'" *RadioFreeEurope*, 30 May 2007, https://www.rferl.org/a/1076805.html.
19. Cyrus Farivar, *The Internet of Elsewhere: The Emergent Effects of a Wired World* (New Brunswick: Rutgers Press, 2011), 137.
20. Sheng Li, "When does internet denial trigger the right of armed self-defense?," *The Yale Journal of International Law*, vol. 38, no. 179 (2013): 200, https://openyls.law.yale.edu/handle/20.500.13051/6651.
21. Andreas Schmidt, "The Estonian cyberattacks," in Jason Healey (ed.), *The Fierce Domain—Conflicts in Cyberspace 1986–2012* (Washington, D.C.: Atlantic Council, 2013), https://www.researchgate.net/publication/264418820_The_Estonian_Cyberattacks.
22. Patrick Kingsley, "How tiny Estonia stepped out of USSR's shadow to become an internet titan," *The Guardian*, 15 April 2012, https://www.theguardian.com/technology/2012/apr/15/estonia-ussr-shadow-internet-titan.
23. S. Haataja, "The 2007 cyber attacks against Estonia and international law on the

24. Andreas Schmidt, "The Estonian Cyberattacks," in Jason Healey (ed.), *The Fierce Domain—Conflicts in Cyberspace 1986–2012* (Washington, D.C.: Atlantic Council, 2013), https://www.researchgate.net/publication/264418820_The_Estonian_Cyberattacks.
25. Monica M. Ruiz,"Is Estonia's approach to cyber defense feasible in the United States?," *War on the Rocks*, 9 January 2018, https://warontherocks.com/2018/01/estonias-approach-cyber-defense-feasible-united-states/.
26. Patrick Kingsley, "How tiny Estonia stepped out of USSR's shadow to become an internet titan," *The Guardian*, 15 April 2012, https://www.theguardian.com/technology/2012/apr/15/estonia-ussr-shadow-internet-titan.
27. Monica M. Ruiz, "Is Estonia's approach to cyber defense feasible in the United States?," *War on the Rocks*, 9 January 2018, https://warontherocks.com/2018/01/estonias-approach-cyber-defense-feasible-united-states/.
28. Elizabeth Schulze, "How a tiny country bordering Russia became one of the most tech-savvy societies in the world," *CNBC*, 8 February 2019, https://www.cnbc.com/2019/02/08/how-estonia-became-a-digital-society.html.
29. Ian Traynor, "Russia accused of unleashing cyberwar to disable Estonia," *The Guardian*, 16 May 2007, https://www.theguardian.com/world/2007/may/17/topstories3.russia.
30. Merle Maigre, "NATO's role in global cyber security," German Marshall Fund, 6 April 2022, https://www.gmfus.org/news/natos-role-global-cyber-security.
31. Stephanie MacLellan and Naomi O'Leary, "Doing battle in cyberspace: How an attack on Estonia changed the rules of the game," *CIGI*, 26 October 2017, https://www.cigionline.org/articles/doing-battle-cyberspace-how-attack-estonia-changed-rules-game/.
32. Hannes Grassegger and Mikael Krogerus, "Fake news and botnets: How Russia weaponised the web," *The Guardian*, 2 December 2017, https://www.theguardian.com/technology/2017/dec/02/fake-news-botnets-how-russia-weaponised-the-web-cyber-attack-estonia.
33. "Our mission & vision," About Us, CCDCOE, https://ccdcoe.org/about-us/.
34. Michael Schmitt, "Tallinn Manual 2.0 on the International Law of Cyber Operations: What it is and isn't," *Just Security*, 9 February 2017, https://www.justsecurity.org/37559/tallinn-manual-2-0-international-law-cyber-operations/.
35. "Centres of Excellence," North Atlantic Treaty Organization, https://www.nato.int/cps/en/natolive/topics_68372.htm.
36. "Four new member states join Estonia-based NATO CCDCOE," *ERR*, 17 May 2023, https://news.err.ee/1608980384/four-new-member-states-join-estonia-based-nato-ccdcoe.
37. Josephine Wolff, "Why Russia hasn't launched major cyber attacks since the invasion of Ukraine," *Time*, 2 March 2022, https://time.com/6153902/russia-major-cyber-attacks-invasion-ukraine/.

38. Merle Maigre, "An e-integration marathon: The potential impact of Ukrainian membership on the EU's digitalisation and cybersecurity," International Centre for Defence and Security, 11 January 2024, https://icds.ee/en/an-e-integration-marathon-the-potential-impact-of-ukrainian-membership-on-the-eus-digitalisation-and-cybersecurity/.
39. Andy Greenberg, "The untold story of NotPetya, the most devastating cyberattack in history," *Wired*, 22 August 2018, https://www.wired.com/story/notpetya-cyberattack-ukraine-russia-code-crashed-the-world/.
40. Josephine Wolff, "How the NotPetya attack is reshaping cyber insurance," *Brookings*, 1 December 2021, https://www.brookings.edu/articles/how-the-notpetya-attack-is-reshaping-cyber-insurance/.
41. Merle Maigre, "An e-integration marathon: The potential impact of Ukrainian membership on the EU's digitalisation and cybersecurity," International Centre for Defence and Security, 11 January 2024, https://icds.ee/en/an-e-integration-marathon-the-potential-impact-of-ukrainian-membership-on-the-eus-digitalisation-and-cybersecurity/.
42. Merle Maigre, "An e-integration marathon: The potential impact of Ukrainian membership on the EU's digitalisation and cybersecurity," International Centre for Defence and Security, 11 January 2024, https://icds.ee/en/an-e-integration-marathon-the-potential-impact-of-ukrainian-membership-on-the-eus-digitalisation-and-cybersecurity/.
43. Merle Maigre, "An e-integration marathon: The potential impact of Ukrainian membership on the EU's digitalisation and cybersecurity," International Centre for Defence and Security, 11 January 2024, https://icds.ee/en/an-e-integration-marathon-the-potential-impact-of-ukrainian-membership-on-the-eus-digitalisation-and-cybersecurity/.
44. Merle Maigre, "An e-integration marathon: The potential impact of Ukrainian membership on the EU's digitalisation and cybersecurity," International Centre for Defence and Security, 11 January 2024, https://icds.ee/en/an-e-integration-marathon-the-potential-impact-of-ukrainian-membership-on-the-eus-digitalisation-and-cybersecurity/.
45. Holli Nelson, "West Virginia, North Carolina Guard join cyber exercise," Air National Guard, 27 April 2022, https://www.ang.af.mil/Media/Article-Display/Article/3011573/west-virginia-north-carolina-guard-join-cyber-exercise/.
46. "World's largest cyber defense exercise Locked Shields kicks off in Tallinn," *Helsinki Times*, 19 April 2023, https://www.helsinkitimes.fi/world-int/23412-world-s-largest-cyber-defense-exercise-locked-shields-kicks-off-in-tallinn.html.
47. M. Kaljurand, "Taking stock of Estonia's multistakeholder cyber diplomacy," in *Building an International Cybersecurity Regime* (Cheltenham, UK: Edward Elgar Publishing 2023), Chapter 11, https://doi.org/10.4337/9781035301546.00018.
48. Parick Tucker and Defense One, "Who will defend tomorrow's digital coun-

tries?," *The Atlantic*, 8 September 2014, https://www.theatlantic.com/international/archive/2014/09/when-a-digital-country-is-in-nato/379806/.
49. Maggie Miller, "How Estonia is helping Ukraine take on Russian cyber threats," *Politico*, 12 August 2022, hhttps://www.politico.com/news/2022/12/07/estonia-ukraine-cybersecurity-russian-hackers-00072925.
50. Peeter Vihma, "Estonia outranks most of the world in Global Cybersecurity Index," *e-Estonia*, 15 June 2022, https://e-estonia.com/estonia-outranks-most-of-the-world-in-global-cybersecurity-index/.
51. Pascale Davies, "Estonia hit by 'most extensive' cyberattack since 2007 amid tensions with Russia over Ukraine war," *Euronews*, 18 August 2022, https://www.euronews.com/next/2022/08/18/estonia-hit-by-most-extensive-cyber-attack-since-2007-amid-tensions-with-russia-over-ukrain.
52. Luukas Ilves: CIO of Estonia January 2022–March 2024. Son of President Toomas Hendrik Ilves.
53. Luukas Ilves (@luukasilves), "Yesterday, Estonia was subject to the most extensive cyber attacks it has faced since 2007. Attempted Ddos attacks targeted both public institutions and the private sector…" Twitter thread, 17 August 2022, https://twitter.com/luukasilves/status/1560105663933587458?ref_src=twsrc%5Etfw%7Ctwcamp%5Etweetembed%7Ctwterm%5E1560105663933587458%7Ctwgr%5E259be909e568a650bf24591e49a899287ff4e34f%7Ctwcon%5Es1_&ref_url=https%3A%2F%2Fwww.euronews.com%2Fnext%2F2022%2F08%2F18%2Festonia-hit-by-most-extensive-cyberattack-since-2007-amid-tensions-with-russia-over-ukrain.
54. "NATO DIANA invests almost €1 million in Estonia-based startups," *ERR*, 12 January 2023, https://news.err.ee/1609182358/nato-diana-invests-almost-1-million-in-estonia-based-startups.
55. Hannes Grassegger and Mikael Krogerus, "Fake news and botnets: How Russia weaponised the web," *The Guardian*, 2 December 2017, https://www.theguardian.com/technology/2017/dec/02/fake-news-botnets-how-russia-weaponised-the-web-cyber-attack-estonia.
56. Cody Slingerlan, "The cost of shutting down the internet," CloudZero Blog, 22 May 2024, https://www.cloudzero.com/blog/cost-of-shutting-down-the-internet/?utm_campaign=etb&utm_medium=newsletter&utm_source=morning_brew.

12. SOFTWARE EATS GOVERNMENT FOR BREAKFAST

1. Chandra Gnanasambandam, Janaki Palaniappan, and Jeremy Schneider, "Every company is a software company: Six 'must dos' to succeed," *McKinsey Quarterly*, 13 December 2022, https://www.mckinsey.com/capabilities/mckinsey-digital/our-insights/every-company-is-a-software-company-six-must-dos-to-succeed
2. Digital public infrastructure refers to online state-created and managed

infrastructure which enables online activities (for example, digital identity or an X-Road style data sharing solution which create a foundation which both the public and private sector can build on to deliver additional services).

3. Toomas Hendrik Ilves, "Oral history interview with Toomas Hendrick Ilves," *Baltic Video Testimonies at Stanford Libraries*, 1 October 2017, video, 51:47, https://purl.stanford.edu/kz740zk2852.

4. "Veterans affairs: Systems modernization, cybersecurity, and IT management issues need to be addressed," Reports & Testimonies, U.S. Government Accountability Office, https://www.gao.gov/products/gao-21-105304.

5. Hettie O'Brien, "The Big Con by Mariana Mazzucato and Rosie Collington review—How consultancy firms cash in," *The Guardian*, 16 February 2023, https://www.theguardian.com/books/2023/feb/16/the-big-con-by-mariana-mazzucato-and-rosie-collington-review-how-consultancy-firms-cash-in#new_tab.

6. John Naughton, "Horrified by Horizon? Then get ready to be totally appalled by AI," *The Guardian*, 13 January 2024, https://www.theguardian.com/uk-news/2024/jan/13/horrified-by-horizon-then-get-ready-to-be-totally-appalled-by-ai.

7. Anna Cooban, "Prison. Bankruptcy. Suicide. How a software glitch and a centuries-old British company ruined lives," *CNN Business*, 13 January 2024, https://www.cnn.com/2024/01/13/business/uk-post-office-fujitsu-horizon-scandal/index.html.

8. Mark Sweeney, "Fixing Horizon bugs would have been too costly, Post Office inquiry told," *The Guardian*, 17 January 2024, https://www.theguardian.com/uk-news/2024/jan/17/post-office-inquiry-fixing-horizon-bugs-fujitsu-developer-gerald-barnes.

9. Mitchell Clark, "Bad ssoftware sent postal workers to jail, because no one wanted to admit it could be wrong," *The Verge*, 23 April 2021, https://www.theverge.com/2021/4/23/22399721/uk-post-office-software-bug-criminal-convictions-overturned.

10. Laura Martin, "Mr Bates vs The Post Office: How a TV drama shook up Britain—in just a week," *BBC*, 12 January 2024, https://www.bbc.com/culture/article/20240112-post-office-scandal-how-a-tv-drama-shook-up-britain-in-just-a-week.

11. Alice Moore, "Post Office Horizon scandal: Four reasons why the government's model for outsourcing is broken," *The Conversation*, 15 January 2024, https://theconversation.com/post-office-horizon-scandal-four-reasons-why-the-governments-model-for-outsourcing-is-broken-220919.

12. Madis Hindre, "Sikkut: Politicians reminded of e-state importance when something breaks," *ERR*, 27 January 2022, https://news.err.ee/1608479921/sikkut-politicians-reminded-of-e-state-importance-when-something-breaks.

13. Maggie Miller, "How Estonia is helping Ukraine take on Russian cyber threats," *Politico*, 12 August 2022, https://www.politico.com/news/2022/12/07/estonia-ukraine-cybersecurity-russian-hackers-00072925.

14. "Estonia directs additional €14.4 million to digital state upgrades in 2022," *ERR*, 6 October 2021, https://news.err.ee/1608361020/estonia-directs-additional-14-4-million-to-digital-state-upgrades-in-2022.
15. M. B. Lapping, "Education in a restoration democracy: The case of Estonia," *Citizenship, Social and Economics Education*, vol. 6, no. 2 (2004), 101–115, https://doi.org/10.2304/csee.2004.6.2.101.
16. Precious failures," *e-Estonia*, 18 August 2021, https://e-estonia.com/precious-failures/.
17. "Despite high-profile failures, government tech is slowly improving," *The Economist*, 18 March 2021, https://www.economist.com/united-states/2021/03/18/despite-high-profile-failures-government-tech-is-slowly-improving.
18. "Information technology: Agencies need to continue addressing critical legacy systems," Reports & Testimonies, U.S. Government Accountability Office, https://www.gao.gov/products/gao-23-106821#:~:text=Fast%20Facts,%24100%20billion%20on%20information%20technology.
19. Greg Godbout, "Scaling proven IT modernization strategies across the federal government," Federation of American Scientists, 22 October 2024, https://fas.org/publication/scaling-proven-it-modernization-strategies-across-the-federal-government/.
20. Meinhard Pulk, "Another crack in the digital state's reputation: Estonia has to apologize to the EU," *Postimees*, 18 November 2022, https://news.postimees.ee/7651636/another-crack-in-the-digital-state-s-reputation-estonia-has-to-apologize-to-the-eu.
21. Aare Lapõnin, "There is software out there," Lapõnin personal blog, 29 September 2018, https://aarelaponin.com/2018/09/29/there-is-software-out-there/.
22. "Culture eats strategy for breakfast," The Management Centre, https://www.managementcentre.co.uk/management-consultancy/culture-eats-strategy-for-breakfast/.

13. INFORMATION AGE

1. "Bye, bye 2022," Articles, *e-Estonia*, 25 January 2023, https://e-estonia.com/bye-bye-2022/.
2. "This month marks a significant milestone for e-Estonia as we celebrate our 15th birthday!...," e-Estonia, Linkedin post, April 2024, https://www.linkedin.com/posts/e-estonia_this-month-marks-a-significant-milestone-activity-7174389301764341760-wn7b?utm_source=share&utm_medium=member_desktop.
3. Jennifer Pahlka, *Recoding America: Why Government Is Failing in the Digital Age and How We Can Do Better* (New York: Metropolitan Press, 2023), 101–102.
4. Jennifer Pahlka, *Recoding America: Why Government Is Failing in the Digital Age and How We Can Do Better* (New York: Metropolitan Press, 2023), 37.
5. Steven M. Teles, "Kludgeocracy in America," *National Affairs*, Fall 2013, https://www.nationalaffairs.com/publications/detail/kludgeocracy-in-america.

6. Peeter Vihma, *Twenty Years of Building Digital Societies: Thinking about the Past and Future of Digital Transformation* (Tallinn: e-Governance Academy, 2023), 11.
7. "Hack Your Bureaucracy: A book by Marina Nitze and Nick Sinai," Hack Your Bureaucracy, https://www.hackyourbureaucracy.com/.
8. Steve Blank, "Why innovation heroes are a sign of a dysfunctional organization," Steve Blank Substack, 20 June 2024, https://steveblank.substack.com/p/why-innovation-heroes-are-a-sign.
9. Jennifer Pahlka, *Recoding America: Why Government Is Failing in the Digital Age and How We Can Do Better* (New York: Metropolitan Press, 2023), 264.
10. Jennifer Pahlka, "The public wants I-95-ness," *Eating Policy*, 27 February 2024, https://eatingpolicy.substack.com/p/the-public-wants-i-95-ness.
11. Felix Richter, "How is the US economy doing after COVID-19?," *World Economic Forum*, 2 February 2023, https://www.weforum.org/agenda/2023/02/us-economy-covid19-inflation/.
12. Information age: Sometimes referred to as the digital age, computer age, or internet age, the term refers to the historical period beginning in the mid to late twentieth century that was characterized by the rise and dominance of information technology throughout both the economy and society.
13. T. Armen Sarkissian, *The Small States Club: How Small Start States Can Save the World* (London: C. Hurst & Co.), Kindle edition, 50.
14. James Dale Davidson and Lord William Rees-Mogg, *The Sovereign Individual: Mastering the Transition to the Information Age (* (New York: Touchstone), Kindle edition, 9.
15. Joshua Keating, "Welcome to the 'neomedieval era,'" *Vox*, 6 February 2024, https://www.vox.com/world-politics/24062198/israel-gaza-middle-east-united-states-war-biden-china-ukraine-putin-russia-taiwan-defense-military.
16. Author interview: Steve Jurvetson, email correspondence, 13 May 2024.
17. Charlie Warzel and Ryan Mac, "These confidential charts show why Facebook bought WhatsApp," *BuzzFeed News*, 5 December 2018, https://www.buzzfeednews.com/article/charliewarzel/why-facebook-bought-whatsapp.
18. Kurt Wagner, "Facebook sees WhatsApp as its future, antitrust suit or not," *Bloomberg*, 9 December 2020, https://www.bloomberg.com/news/features/2020-12-09/facebook-fb-plans-to-turn-messaging-app-whatsapp-into-a-moneymaking-business?embedded-checkout=true.
19. Robert McMillan, "You may not use WhatsApp, but the rest of the world sure does," *Wired*, 20 February 2014, https://www.wired.com/2014/02/whatsapp-rules-rest-world/.
20. Akshat Rathi, "WhatsApp bought for $19 billion, what do its employees get?," *The Conversation*, 20 February 2014, https://theconversation.com/whatsapp-bought-for-19-billion-what-do-its-employees-get-23496.
21. Roland Martin, "WhatsApp," *Britannica*, https://www.britannica.com/topic/WhatsApp.
22. Paolo Confino, "Could AI create a one-person unicorn? Sam Altman thinks so—And Silicon Valley sees the technology 'waiting for us,'" *Fortune*, 4 February

2024, https://fortune.com/2024/02/04/sam-altman-one-person-unicorn-silicon-valley-founder-myth/amp/.

23. Zak Kukoff (@zck), "Prediction: A team of fewer than 5 people will make a move that grosses >$50M at box office using text to video models and non-union (ie non-WGA, SAG, etc) labor within 5 years," Twitter post, 15 February 2024, https://twitter.com/zck/status/1758232504857690565?s=46&t=IzPrVbo2SjIuAkI3DMvhxw.

24. Eleanor Ainge Roy, "New Zealand gave Pether Thiel citizenship after he spent just 12 days there," *The Guardian*, 29 June 2017, https://www.theguardian.com/world/2017/jun/29/new-zealand-gave-peter-thiel-citizenship-after-spending-just-12-days-there.

25. Naomi Buchanan, "OpenAI and Google CEOs among America's tech moguls acquiring more visas, citizenships," *Investopedia*, 9 September 2023, https://www.investopedia.com/openai-and-google-ceos-among-american-tech-moguls-buying-more-visas-and-citizenships-7966045.

26. Theodore Schleifer, "The former CEO of Google has applied to become a citizen of Cyprus," *Vox*, 9 November 2020, https://www.vox.com/recode/2020/11/9/21547055/eric-schmidt-google-citizen-cyprus-european-union.

27. Alice Kantor, "Europe's golden visas are booming, despite calls to get rid of them," *Bloomberg*, 15 August 2023, https://www.bloomberg.com/news/articles/2023-08-15/europe-s-golden-visas-are-booming-despite-calls-to-get-rid-of-them.

28. Robert Frank, "The rich are getting second passports, citing risk of instability," *CNBC*, 10 April 2024, https://www.cnbc.com/2024/04/10/rich-americans-get-second-passports-citing-risk-of-instability.html?utm_source=pocket-newtab-en-us.

29. Kim Parker, Juliana Menasce Horowitz, and Rachel Minkin, "COVID-19 pandemic continues to reshare work in America," *Pew Research Center*, 16 February 2022, https://www.pewresearch.org/social-trends/2022/02/16/covid-19-pandemic-continues-to-reshape-work-in-america/.

30. Emma Agyemang, "Countries wooing corporate digital nomads hope to make them stay," *Financial Times*, 18 May 2024, https://www.ft.com/content/6fa06cf8-1724-4b0d-a3c2-59dcbafbdebd?utm_source=substack&utm_medium=email.

31. Alejandra O'Connell-Domenech, "Record number of Americans granted temporary resident visas to live in Mexico," *The Hill*, 7 November 2022, https://thehill.com/changing-america/sustainability/infrastructure/3723574-record-number-of-americans-granted-temporary-resident-visas-to-live-in-mexico/.

32. Ward Williams, "Countries offering digital nomad visas," *Investopedia*, https://www.investopedia.com/countries-offering-digital-nomad-visas-5190861.

33. "12 month Barbados Welcome Stamp," PWC, https://www.pwc.com/bb/en/services/pdf/12-month-barbados-welcome-stamp.pdf.

34. Emma Agyemang, "Countries wooing corporate digital nomads hope to make

them stay," *Financial Times*, 18 May 2024, https://www.ft.com/content/6fa06cf8-1724-4b0d-a3c2-59dcbafbdebd?utm_source=substack&utm_medium=email.

35. Emma Agyemang, "Countries wooing corporate digital nomads hope to make them stay," *Financial Times*, 18 May 2024, https://www.ft.com/content/6fa06cf8-1724-4b0d-a3c2-59dcbafbdebd?utm_source=substack&utm_medium=email.

36. Jay Peters, "Amazon is making its employees come back to the office five days a week," *The Verge*, 16 September 2024, https://www.theverge.com/2024/9/16/24246428/amazon-return-to-office-five-days-a-week.

37. Emma Goldberg, "Can remote workers reverse brain drain?," *The New York Times*, 16 October 2024, https://www.nytimes.com/2024/10/16/business/tulsa-remote-workers.html.

38. Karoli Hindriks, "Enter the era of the digital nomad," *Sifted*, 15 June 2020, https://sifted.eu/articles/digital-nomad-visa-estonia.

39. Rosie Spinks, "Estonia is launching a digital nomad visa," *Quartz*, 27 February 2018, https://qz.com/quartzy/1216964/estonia-is-launching-a-digital-nomad-visa.

40. Hannah Brown, "FAQs about Estonia's Digital Nomad Visa," E-Residency Blog, 7 July 2020, https://medium.com/e-residency-blog/faqs-about-estonias-digital-nomad-visa-b04f12551e30.

41. "Minister suggests lowering Estonia's requirements for digital nomads," *ERR*, 27 February 2024, https://news.err.ee/1609265727/minister-suggests-lowering-estonia-s-requirements-for-digital-nomads.

42. Emma Agyemang, "Countries wooing corporate digital nomads hope to make them stay," *Financial Times*, 18 May 2024, https://www.ft.com/content/6fa06cf8-1724-4b0d-a3c2-59dcbafbdebd?utm_source=substack&utm_medium=email.

43. Dave Cook, "Remote working: how a surge in digital nomads is pricing out local communities around the world," *The Conversation*, 31 March 2023, https://theconversation.com/remote-working-how-a-surge-in-digital-nomads-is-pricing-out-local-communities-around-the-world-200670.

44. Joshua Keating, *Invisible Countries: Journeys to the Edge of Nationhood* (New Haven and London: Yale University Press, 2018), Kindle edition, 200.

45. Marissa Newman, "Ex-Google CEO Schmidt urges US army to replace tanks with drones," *Bloomberg*, 30 October 2024, https://www.bloomberg.com/news/articles/2024-10-30/ex-google-ceo-schmidt-urges-us-army-to-replace-tanks-with-drones?srnd=phx-technology.

46. Sarah Emerson and Richard Nieva, "Eric Schmidt's secret military project revealed: Attack drones," *Forbes*, 6 February 2024, https://www.forbes.com/sites/sarahemerson/2024/01/23/eric-schmidts-secret-white-stork-project-aims-to-build-ai-combat-drones/.

47. J. M. Page and J. Williams, "Drones, Afghanistan, and beyond: Towards anal-

ysis and assessment in context," *European Journal of International Security*, vol. 7, no. 3 (2022), 283–303, https://doi.org/10.1017/eis.2021.19.

48. Molly Dunigan, "A Lesson from Iraq War: How to outsource war to private contractors," *Rand*, 19 March 2013, https://www.rand.org/pubs/commentary/2013/03/a-lesson-from-iraq-war-how-to-outsource-war-to-private.html.

49. Lauren Feiner, "Startup investors are fueling a boom in U.S. defense tech as China standoff opens doors at home," *CNBC*, 7 September 2023, https://www.cnbc.com/2023/09/07/startup-investors-fuel-boom-in-us-defense-tech-amid-china-standoff.html.

50. Shaan Shaikh and Wes Rumbaugh, "The air and missile war in Nagorno-Karabakh: Lessons for the future of strike and defense," *CSIS*, 8 December 2020, https://www.csis.org/analysis/air-and-missile-war-nagorno-karabakh-lessons-future-strike-and-defense.

51. Antonio Cascais, "Mercenary armies in Africa," *Deutsche Welle*, 15 April 2022, https://www.dw.com/en/the-rise-of-mercenary-armies-in-africa/a-61485270.

52. Admiral James Stavridis, "Admiral James Stavridis: Pre-war America, the U.S. navy, and how the lessons of the 1930s apply to today," interview by Marshall Kosloff, The Realignment Podcast, 8 October 2024, audio, 10:05, https://therealignment.simplecast.com/episodes/511-stavridis.

53. Paul Blumenthal, "The Biden Administration is making it easier to get government benefits," *HuffPost*, 16 September 2023, https://www.huffpost.com/entry/biden-government-benefits_n_65030635e4b01f6f9b9baeee?ari.

54. "Taxing Wages—The United States," Taxing Wages 2024, OECD, https://www.oecd.org/tax/tax-policy/taxing-wages-united-states.pdf.

55. Jane Edwards, "US Chamber of Commerce report: Government digitization could generate $1t annually," *ExecutiveGov*, 19 October 2022, https://executivegov.com/2022/10/us-chamber-of-commerce-government-digitization-could-generate-1t-annually/.

56. "Government digitization: Transforming government to better serve Americans," U.S. Chamber of Commerce, 17 October 2022, https://www.uschamber.com/technology/government-digitization-transforming-government-to-better-serve-americans#:~:text=Reporting%20suggest%20that%20government%20digitization,build%20resilience%2C%20and%20eliminate%20waste.

57. Gary Hamel and Michele Zanini, "Excess management is costing the U.S. $3 trillion per year," *Harvard Business Review*, 5 September 2016, https://hbr.org/2016/09/excess-management-is-costing-the-us-3-trillion-per-year.

58. Billy Mitchell, "Clare Martorana and Raylene Yung call for continued congressional support for TMF," *Fedscoop*, 18 April 2023, https://fedscoop.com/federal-cio-and-raylene-yung-call-for-continued-support-for-tmf/.

59. "Release: Rep. Khanna on the passage of the 21st century Integrated Digital Experience Act (IDEA)," Press Releases, Congressman Ro Khanna, 29 November 2018, https://khanna.house.gov/media/press-releases/release-rep-khanna-passage-21st-century-integrated-digital-experience-act-idea.

NOTES

60. Nihal Krishan, "Rep. Ro Khanna calls for accountability of federal leaders on digital services modernization," *Fedscoop*, 21 November 2023, https://fedscoop.com/rep-ro-khanna-calls-for-accountability-of-federal-leaders-on-digital-services-modernization/.

14. COUNTRY-AS-A-SERVICE

1. Emma Agyemang, "Countries wooing corporate digital nomads hope to make them stay," *Financial Times*, 18 May 2024, https://www.ft.com/content/6fa06cf8-17244b0d-a3c2-59dcbafbdebd?utm_source=substack&utm_medium=email.
2. Taavi Kotka, "Estonia is demonstrating how government should work in a digital world," *Vice*, 10 June 2016, https://www.vice.com/en/article/pgk3gg/estonia-is-demonstrating-how-government-should-work-in-a-digital-world.
3. Holger Roonemaa, "Tim Draper: My heroes were Gorbachev, Washington and Deng. Now the Estonian President and Prime Minister are on the list too!," *Life in Estonia*, June 2016, https://investinestonia.com/tim-draper-my-heroes-were-gorbachev-washington-and-deng-now-the-estonian-president-and-prime-minister-are-on-the-list-too/.
4. Jef Feely, "Musk slams Delaware, hints at Tesla Texas move after pay snub," *Bloomberg*, 30 January 2024, https://www.bloomberg.com/news/articles/2024-01-31/musk-disses-tiny-delaware-after-its-court-zaps-his-mega-pay-plan?sref=KkPzpZvz&srnd=premium.
5. "Our economy," Government of the Virgin Islands, https://bvi.gov.vg/content/our-economy.
6. Edwin Wee, "Stripe Atlas: The first five years and 20,000 startups," Stripe Corporate Blog, 1 July 2021, https://stripe.com/blog/atlas-first-five-years.
7. "Stripe's 2023 annual letter," *Stripe*, https://stripe.com/annual-updates/2023.
8. Edwin Wee, "Stripe Atlas: The first five years and 20,000 startups," Stripe Corporate Blog, 1 July 2021, https://stripe.com/blog/atlas-first-five-years.
9. Felix Richter, "Charted: There are more mobile phones than people in the world," *World Economic Forum*, 11 April 2023, https://www.weforum.org/agenda/2023/04/charted-there-are-more-phones-than-people-in-the-world/.
10. "The Henley Passport Index," Henley & Partners, https://www.henleyglobal.com/passport-index/ranking.
11. David Eaves, "The digital systems every country needs to succeed in the 21st century," Harvard Kennedy School, 25 May 2022, https://ash.harvard.edu/digital-systems-every-country-needs-succeed-21st-century.
12. Taavi Kotka, "Country as a service: Estonia's new model," *e-Estonia*, 19 May 2016, https://e-estonia.com/country-as-a-service-estonias-new-model/.
13. Taavi Kotka, "Country as a service: Estonia's new model," *e-Estonia*, 19 May 2016, https://e-estonia.com/country-as-a-service-estonias-new-model/.
14. Siim Sikkut, *Digital Government Excellence: Lessons From Effective Digital Leaders* (Hoboken: John Wiley & Sons 2022), 241–242.

15. Guillaume Long and Alexander Main, "How a start-up utopia became a nightmare for Honduras," *Foreign Policy*, 24 January 2024, https://foreignpolicy.com/2024/01/24/honduras-zedes-us-prospera-world-bank-biden-castro/.
16. Gustavo Palencia, "Honduras top court declares self-governing ZEDE zones unconstitutional," *Reuters*, 20 September 2024, https://www.reuters.com/world/americas/honduras-top-court-declares-self-governing-zede-zones-unconstitutional-2024-09-20/.
17. Stephen Alpher, "Praxis raises $15M in Series A funding round," *Coindesk*, 3 March 2022, https://www.coindesk.com/business/2022/03/03/paradigm-among-investors-as-praxis-raises-15m-in-series-a-funding-round/.
18. Holger Roonemaa, "Tim Draper: My heroes were Gorbachev, Washington and Deng. Now the Estonian President and Prime Minister are on the list too!," *Life in Estonia*, June 2016, https://investinestonia.com/tim-draper-my-heroes-were-gorbachev-washington-and-deng-now-the-estonian-president-and-prime-minister-are-on-the-list-too/.

15. TEN MILLION ESTONIANS

1. "Pope Francis becomes an Estonian e-resident," *Estonian World*, 25 September 2018, https://estonianworld.com/business/pope-francis-becomes-an-estonian-e-resident/.
2. Silver Tambur, "Angela Merkel becomes Estonian e-resident," *Estonian World*, 25 August 2016, https://estonianworld.com/technology/angela-merkel-becomes-estonian-e-resident/.
3. April Rinne, "One of Estonia's first 'e-residents' explains what it means to have digital citizenship," *Quartz*, 1 April 2018, https://qz.com/work/1241833/one-of-estonias-first-e-residents-explains-what-it-means-to-have-digital-citizenship.
4. Disclosure: I served as Head of Business Development for e-Residency in 2018–2019 and worked directly with many of the individuals mentioned throughout the section.
5. Arielle Pardes, "Estonia's e-Residency program is the future of immigration," *Vice*, 5 May 2016, https://www.vice.com/en/article/avyx5a/estonias-e-residency-program-is-the-future-of-immigration.
6. Silver Tambur, "Estonia now has 100,000 digital residents," *Estonian World*, February 22, 2023, https://estonianworld.com/business/estonia-now-has-100000-digital-residents/.
7. Rob Garver, "Estonia offers 'electronic residency' worldwide," *CNBC*, 2 December 2014, https://www.cnbc.com/2014/12/02/estonia-offers-electronic-residency-worldwide.html.
8. Matt Reynolds, "'Land is so yesterday': E-residents and 'digital embassies' could replace country borders," *Wired*, 17 October 2016, https://www.wired.co.uk/article/taavi-kotka-estonian-government.
9. Silver Tambur, "Estonia becomes the first country in the world to offer

e-residency," *Estonian World*, 1 December 2014, https://estonianworld.com/business/estonia-becomes-first-country-world-offer-e-residency/.

10. Maeve Shearlaw, "A Brexit bolthole? For €100 you can become an e-resident of an EU country you've never visited," *The Guardian*, 15 September 2016, https://www.theguardian.com/world/2016/sep/15/estonia-e-residency-european-union-brexit-eu-referendum.
11. Andrus Viirg, "Oral history interview," interview by Anders Hjemdahl and Camilla Andersson, Stanford Digital Repository, 30 September 2017, https://purl.stanford.edu/mf560sv0095.
12. Arielle Pardes, "Estonia's e-Residency program is the future of immigration," *Vice*, 5 May 2016, https://www.vice.com/en/article/avyx5a/estonias-e-residency-program-is-the-future-of-immigration.
13. Bartosz Chmielewski, "How to postpone a demographic crisis. Estonia and the lifeline of immigration," *OSW Centre for Eastern Studies*, 17 August 2023, https://www.osw.waw.pl/en/publikacje/osw-commentary/2023-08-17/how-to-postpone-a-demographic-crisis-estonia-and-lifeline.
14. Arielle Pardes, "Estonia's e-Residency program is the future of immigration," *Vice*, 5 May 2016, https://www.vice.com/en/article/avyx5a/estonias-e-residency-program-is-the-future-of-immigration.
15. T. Armen Sarkissian, *The Small States Club: How Small Start States Can Save the World* (London: C. Hurst & Co., 2023), Kindle edition, 182.
16. Silver Tambur, "Estonia becomes the first country in the world to offer e-residency," *Estonian World*, 1 December 2014, https://estonianworld.com/business/estonia-becomes-first-country-world-offer-e-residency/.
17. Arielle Pardes, "Estonia's e-Residency program is the future of immigration," *Vice*, 5 May 2016, https://www.vice.com/en/article/avyx5a/estonias-e-residency-program-is-the-future-of-immigration.
18. Agaate Antson, "E-residency creator: It is Estonia's soft power," *Postimees*, 25 November 2019, https://news.postimees.ee/6834270/e-residency-creator-it-is-estonia-s-soft-power.
19. "Testimony of Mr. Edward Lucas," Senate Foreign Relations Committee Subcommittee on European Affairs, 8 July 2014, https://www.foreign.senate.gov/imo/media/doc/Lucas_Testimony1.pdf.
20. Uri Friedman, "The world now has its first e-resident," *The Atlantic*, 1 December 2024, https://www.theatlantic.com/international/archive/2014/12/the-world-has-its-first-e-resident/383277/.
21. "Journalist Edward Lucas becomes the 1st Estonian e-resident," *ERR*, 1 December 2014, https://news.err.ee/114443/journalist-edward-lucas-becomes-the-1st-estonian-e-resident.
22. Mamie Joeveer, "Estonia's rise as a high-tech leader boils down to one notion: Think globally from the start," *Forbes*, 31 December 2014, https://www.forbes.com/sites/mamiejoeveer/2014/12/31/estonias-rise-as-a-high-tech-leader-boils-down-to-one-notion-think-globally-from-the-start/?sh=13d69c785b8d.

23. Dan Primack, "DFJ to become Threshold Ventures," *Axios*, 16 January 2019, https://www.axios.com/2019/01/16/draper-fisher-jurvetson-name-change-threshold-ventures.
24. Cyrus Farivar, "Estonia wants to give us all digital ID cards, make us 'e-residents,'" *Ars Technica*, 8 December 2014, https://arstechnica.com/tech-policy/2014/12/estonia-wants-to-give-us-all-digital-id-cards-make-us-e-residents/.
25. Ede Schank Tamkivi, "e-Residency: the success story of building a digital nation," *Life in Estonia*, January 2020, https://investinestonia.com/e-residency-the-success-story-of-building-a-digital-nation/.
26. Silver Tambur, "Estonia becomes the first country in the world to offer e-Residency," *Estonian World*, 1 December 2014, https://estonianworld.com/business/estonia-becomes-first-country-world-offer-e-residency/.
27. Kaspar Korjus, "E-Residency is 4 years old so here's 4 surprising facts about the programme," E-Residency Blog, 30 November 2018, https://medium.com/e-residency-blog/e-residency-is-4-years-old-so-heres-4-surprising-facts-about-the-programme-c3a9d64c988d.
28. April Rinne, "One of Estonia's first 'e-residents' explains what it means to have digital citizenship," *Quartz*, 1 April 2018, https://qz.com/work/1241833/one-of-estonias-first-e-residents-explains-what-it-means-to-have-digital-citizenship.
29. Daniela Godoy, "E-Residency joins forces with the UN to empower entrepreneurs in the developing world," Estonia.ee, October 2017, https://estonia.ee/e-residency-joins-forces-with-the-un-to-empower-entrepreneurs-in-the-developing-world/.
30. Maeve Shearlaw, "A Brexit bolthole? For €100 you can become an e-resident of an EU country you've never visited," *The Guardian*, 15 September 2016, https://www.theguardian.com/world/2016/sep/15/estonia-e-residency-european-union-brexit-eu-referendum.
31. Trevor Clawson, "Beyond Brexit: Why Estonia's eResidency scheme attracts E.U. startups," *Forbes*, 8 January 2024, https://www.forbes.com/sites/trevorclawson/2023/12/26/beyond-brexit-why-estonias-e-residency-scheme-attracts-eu-startups/.
32. Nathan Heller, "Estonia, the digital republic," *The New Yorker*, 11 December 2017, https://www.newyorker.com/magazine/2017/12/18/estonia-the-digital-republic.
33. Hannah Brown, "A first step towards mobile e-Residency," e-Residency Blog, 25 September 2024, https://www.e-resident.gov.ee/blog/posts/a-first-step-towards-mobile-e-residency/.
34. "Estonia's e-Residency program contributes over €200 million to state budget," *ERR*, 1 December 2023, https://news.err.ee/1609182058/estonia-s-e-residency-program-contributes-over-200-million-to-state-budget.
35. "Estonia: The number of e-residents overtakes the population of Tartu," *The*

Baltic Times, 13 February 2023, https://www.baltictimes.com/estonia__the_number_of_e-residents_overtakes_the_population_of_tartu/.

36. Ede Schank Tamkivi, "E-Residency: The success story of building a digital nation," *Life in Estonia*, January 2020, https://investinestonia.com/e-residency-the-success-story-of-building-a-digital-nation/.
37. "Estonian E-Residency," Sten Tamkivi Personal Website, last updated June 9, 2020, https://tamkivi.com/project/e-residency/.
38. Craig Turp-Balazs, "E-residency: Estonia's government start-up," *Emerging Europe*, 19 March 2020, https://emerging-europe.com/news/e-residency-estonias-government-start-up/.
39. Maya Middlemiss, "Notable e-residents of Estonia," *Xolo*, 21 April 2023, https://blog.xolo.io/notable-e-residents-of-estonia.
40. Agaate Antson, "E-residency creator: It is Estonia's soft power," Postimees, 25 November 2019, https://news.postimees.ee/6834270/e-residency-creator-it-is-estonia-s-soft-power.
41. "First ever e-Residency card of Azerbaijan presented to the Head of the EU Delegation," *European Union External Action*, 26 October 2018, https://www.eeas.europa.eu/node/52823_en.
42. "Lithuania launches own national e-residency scheme," *ERR*, 2 January 2021, https://news.err.ee/1225954/lithuania-launches-own-national-e-residency-scheme.
43. Merle Maigre, "An e-integration marathon: The potential impact of Ukrainian membership on the EU's digitalisation and cybersecurity," *International Centre for Defence and Security*, 11 January 2024, https://icds.ee/en/an-e-integration-marathon-the-potential-impact-of-ukrainian-membership-on-the-eus-digitalisation-and-cybersecurity/.
44. Matti Ylonen, Wolfgang Drechsler, and Veiko Lember, "Online incorporation platforms in Estonia and beyond: How administrative spillover effects hamper international taxation," *Transnational Corporations Journal*, vol. 30, no. 1 (April 2023), https://ssrn.com/abstract=4431829.
45. Matti Ylonen, Wolfgang Drechsler, and Veiko Lember, "Online incorporation platforms in Estonia and beyond: How administrative spillover effects hamper international taxation," *Transnational Corporations Journal*, vol. 30, no. 1 (April 2023), https://ssrn.com/abstract=4431829.
46. Matti Ylonen, Wolfgang Drechsler, and Veiko Lember, "Online incorporation platforms in Estonia and beyond: How administrative spillover effects hamper international taxation," *Transnational Corporations Journal*, vol. 30, no. 1 (April 2023), https://ssrn.com/abstract=4431829.
47. Kalev Aasmae, "Estonia's new e-residency security focus: 'You can't launder money with a digital ID,'" *ZDNet*, 9 July 2019, https://www.zdnet.com/article/estonias-new-e-residency-security-focus-you-cant-launder-money-with-a-digital-id/.
48. Mohammad Musharraf, "Crypto scammers plague Estonia's e-residency pro-

gram," *Cointelegraph*, 28 September 2020, https://cointelegraph.com/news/crypto-scammers-plague-estonias-e-residency-program.

49. "Council of Europe report strongly criticizes Estonian e-residency program," *ERR*, 27 January 2023, https://news.err.ee/1608865556/council-of-europe-report-strongly-criticizes-estonian-e-residency-program.

50. Trevor Clawson, "Beyond Brexit: Why Estonia's eResidency scheme attracts E.U. startups," *Forbes*, 26 December 2023, https://www.forbes.com/sites/trevorclawson/2023/12/26/beyond-brexit-why-estonias-e-residency-scheme-attracts-eu-startups/?sh=7b41055c535a.

51. Taavi Kotka, "Estonia is demonstrating how government should work in a digital world," *Vice*, 10 June 2016, https://www.vice.com/en/article/pgk3gg/estonia-is-demonstrating-how-government-should-work-in-a-digital-world.

52. Author's note: it is my experience having worked for the program that while bureaucracy is a major issue that drives people to use e-Residency, there are certainly many, especially from Western Europe, who find the allure of a potential lower tax environment like Estonia to be one of the main value propositions of the program.

53. Rainer Kattel and Ines Mergel, "Is Estonia the Silicon Valley of digital government?," UCL Institute for Innovation and Public Purpose Blog, 28 September 2018, https://medium.com/iipp-blog/is-estonia-the-silicon-valley-of-digital-government-bf15adc8e1ea.

54. Kalle Palling, "Country as a service (Estonian experience)," YouTube, video, 8:51, https://www.youtube.com/watch?v=W5rAc5B3Ggg.

55. Agaate Antson, "E-residency creator: It is Estonia's soft power," *Postimees*, 25 November 2019, https://news.postimees.ee/6834270/e-residency-creator-it-is-estonia-s-soft-power.

56. Translated transcript derived from an interview on 23 November 2021 with Taavi Kotka, https://searchworks.stanford.edu/view/mw533gy5707.

57. Kaspar Korjus, "We're planning to launch estcoin—and that's only the start," *Medium*, 19 December 2017, https://medium.com/@kaspar.korjus/were-planning-to-launch-estcoin-and-that-s-only-the-start-310aba7f3790.

58. Ryan Browne, "Estonia says it won't issue a national cryptocurrency and never planned to," *CNBC*, 4 June 2018, https://www.cnbc.com/2018/06/04/estonia-wont-issue-national-cryptocurrency-estcoin-never-planned-to.html.

59. Kaspar Korjus, "We're planning to launch estcoin—and that's only the start," *Medium*, 19 December 2017, https://medium.com/@kaspar.korjus/were-planning-to-launch-estcoin-and-that-s-only-the-start-310aba7f3790.

60. Matt Reynolds, "'Land is so yesterday': E-residents and 'digital embassies' could replace country borders," *Wired*, 17 October 2016, https://www.wired.co.uk/article/taavi-kotka-estonian-government.

61. "Estonia's e-Residency program contributes over €200 to state budget," *ERR*, 1 December 2023, https://news.err.ee/1609182058/estonia-s-e-residency-program-contributes-over-200-million-to-state-budget.

62. "B.EST Solutions helps launch m-Residency in Azerbaijan," B.EST Solutions,

30 July 2020, https://bestsolutions.ee/2020/07/b-est-solutions-helps-launch-m-residency-in-azerbaijan/.
63. Nathan Heller, "Estonia, the digital republic," *The New Yorker*, 11 December 2017, https://www.newyorker.com/magazine/2017/12/18/estonia-the-digital-republic.

16. CLOUD COUNTRY

1. Don Moynihan and Gulsanna Mamediieva, "Digital resilience in Ukraine," *Can We Still Govern?*, 31 October 2023, https://donmoynihan.substack.com/p/digital-resilience-in-ukraine.
2. "Corruption Perceptions Index: Ukraine 2019," Transparency International, https://www.transparency.org/en/cpi/2019/index/ukr.
3. "Mykhailo Fedorov," Government Portal Official Website, https://www.kmu.gov.ua/en/profile/mikhaylo-fedorov.
4. "Jaanika Merilo begins work as Estonia's e-Health strategy manager," *ERR*, 3 January 2024, https://news.err.ee/1609210702/jaanika-merilo-begins-work-as-estonia-s-e-health-strategy-manager.
5. Jaanika Merilo, "What we can learn from digital Ukraine?," interview by Florian Marcus, *e-Estonia*, 2022, video, 7:50, https://www.youtube.com/watch?v=HyvwyzVMWR0.
6. Jaanika Merilo, "What we can learn from digital Ukraine?," interview by Florian Marcus, *e-Estonia*, 2022, video, 7:50, https://www.youtube.com/watch?v=HyvwyzVMWR0.
7. Mykhailo Fedorov, "Ukraine's digital revolution is gaining momentum," *Atlantic Council*, 7 September 2021, https://www.atlanticcouncil.org/blogs/ukrainealert/ukraines-digital-revolution-is-gaining-momentum/.
8. "Ministry of Digital Transformation launches e-residency program in closed beta testing mode," *interfax Ukraine*, 27 December 2023, https://en.interfax.com.ua/news/economic/956742.html.
9. "European Union supports the launch of uResidency: Foreigners will be able to become Ukrainian e-residents," Delegation of the European Union to Ukraine, https://www.eeas.europa.eu/delegations/ukraine/european-union-supports-launch-uresidency-foreigners-will-be-able-become-ukrainian-e-residents_en.
10. Mykhailo Fedorov, "Ukraine's digital revolution is gaining momentum," *Atlantic Council*, 7 September 2021, https://www.atlanticcouncil.org/blogs/ukrainealert/ukraines-digital-revolution-is-gaining-momentum/.
11. Kateryna Denisova, "Ukraine launches online marriages to unite couples separated by war," *Kyiv Independent*, 23 September 2024, https://kyivindependent.com/it-looked-unusual-we-risked-how-ukrainians-getting-married-online/.
12. "Diia," The Best Inventions of 2024, *Time*, https://time.com/7094556/diia/.
13. The Ministry of Digital Transformation, "Sharing the experience of digital Ukraine: first results of cooperation with Ecuador…," LinkedIn Post, 5 June

2024, https://www.linkedin.com/feed/update/urn:li:activity:7204088649569779713/.
14. James Andrew Lewis, "Cyber war and Ukraine," *CSIS*, 16 June 2022, https://www.csis.org/analysis/cyber-war-and-ukraine#:~:text=February%202022%3A%20Hackers%20deployed%20a,a%20Russian%20GRU%2Daffiliated%20group.
15. Ryan White, "How the cloud saved Ukraine's data from Russian attacks," *C4ISRNET*, 22 June 2022, https://www.c4isrnet.com/2022/06/22/how-the-cloud-saved-ukraines-data-from-russian-attacks/.
16. Russ Mitchell, "How Amazon put Ukraine's 'government in a box'—And saved its economy from Russia," *Los Angeles Times*, 15 December 2022, https://www.latimes.com/business/story/2022-12-15/amazon-ukraine-war-cloud-data.
17. Russ Mitchell, "How Amazon put Ukraine's 'government in a box'—And saved its economy from Russia," *Los Angeles Times*, 15 December 2022, https://www.latimes.com/business/story/2022-12-15/amazon-ukraine-war-cloud-data.
18. Katherine Tangalakis-Lippert, "Amazon helped rescue the Ukrainian government and economy using suitcase-sized hard drives brought over the Polish border: 'You can't take out the cloud with a cruise missile,'" *Business Insider*, 18 December 2022, https://www.businessinsider.com/amazon-saved-the-ukrainian-government-with-suitcase-sized-hard-drives-2022–12.
19. Brad Smith, "Defending Ukraine: Early lessons from the cyber war," Microsoft Blog, 22 June 2022, https://blogs.microsoft.com/on-the-issues/2022/06/22/defending-ukraine-early-lessons-from-the-cyber-war/.
20. Katherine Tangalakis-Lippert, "Amazon helped rescue the Ukrainian government and economy using suitcase-sized hard drives brought over the Polish border: 'You can't take out the cloud with a cruise missile,'" *Business Insider*, 18 December 2022, https://www.businessinsider.com/amazon-saved-the-ukrainian-government-with-suitcase-sized-hard-drives-2022–12.
21. Katherine Tangalakis-Lippert, "Amazon helped rescue the Ukrainian government and economy using suitcase-sized hard drives brought over the Polish border: 'You can't take out the cloud with a cruise missile,'" *Business Insider*, 18 December 2022, https://www.businessinsider.com/amazon-saved-the-ukrainian-government-with-suitcase-sized-hard-drives-2022–12.
22. Beatric Nolan, "Zelenskyy awards Amazon the Ukraine peace prize after AWS helped save its 'digital infrastructure,'" *Business Insider*, 6 July 2022, https://www.businessinsider.com/zelenskyy-amazon-ukraine-peace-prize-digital-war-support-aws-2022–7.
23. "Safeguarding Ukraine's data to preserve its present and build its future," AWS, Amazon Corporate Blog, 9 June 2022, https://www.aboutamazon.com/news/aws/safeguarding-ukraines-data-to-preserve-its-present-and-build-its-future.
24. Don Moynihan and Gulsanna Mamediieva, "Digital resilience in Ukraine," *Can We Still Govern?*, 31 October 2023, https://donmoynihan.substack.com/p/digital-resilience-in-ukraine.

25. "eVorog—Chatbot for civil intelligence in Telegram," YouTube, https://www.youtube.com/watch?v=VTAxA3Omfmw.
26. "How a chatbot has turned Ukrainian civilians into digital resistance fighters," *The Economist*, 22 February 2023, https://www.economist.com/the-economist-explains/2023/02/22/how-a-chatbot-has-turned-ukrainian-civilians-into-digital-resistance-fighters.
27. Kyle Hiebert, "The reconstruction of Ukraine can inform the West's digital transformation," Centre for International Governance Innovation, 21 August 2023, https://www.cigionline.org/articles/the-reconstruction-of-ukraine-can-inform-the-wests-digital-transformation/.
28. "Safeguarding Ukraine's data to preserve its present and build its future," AWS, Amazon Corporate Blog, 9 June 2022, https://www.aboutamazon.com/news/aws/safeguarding-ukraines-data-to-preserve-its-present-and-build-its-future.
29. Richard Martyn-Hemphill, "'Old Cold War warrior' imagines war with Russia," *Politico*, 6 June 2016, https://www.politico.eu/article/old-cold-war-warrior-imagines-war-with-russia-nato-brexit-vladimir-putin/.
30. Yuliya Talmazan, "Data security meets diplomacy: Why Estonia is storing its data in Luxembourg," *NBC News*, 25 June 2019, https://www.nbcnews.com/news/world/data-security-meets-diplomacy-why-estonia-storing-its-data-luxembourg-n1018171.
31. Nick Robinson, Laura Kask, and Robert Krimmer, "The Estonian data embassy and the applicability of the Vienna Convention: An exploratory analysis," *ICEGOV '19: Proceedings of the 12th International Conference on Theory and Practice of Electronic Governance* (2018), https://doi.org/10.1145/3326365.3326417.
32. Yuliya Talmazan, "Data security meets diplomacy: Why Estonia is storing its data in Luxembourg," NBC News, 25 June 2019, https://www.nbcnews.com/news/world/data-security-meets-diplomacy-why-estonia-storing-its-data-luxembourg-n1018171.
33. "E-embassies ensure IT Security and diplomatic protection," E-Embassies in Luxembourg, Luxembourg, https://luxembourg.public.lu/en/invest/innovation/e-embassies-in-luxembourg.html.
34. Russ Mitchell, How Amazon put Ukraine's 'government in a box'—And saved its economy from Russia," *Los Angeles Times*, 15 December 2022, https://www.latimes.com/business/story/2022-12-15/amazon-ukraine-war-cloud-data.
35. "Establishing the first Data Embassy in the world," Case Study Library, OECD Observatory of Public Sector Innovation, https://oecd-opsi.org/innovations/establishing-the-first-data-embassy-in-the-world/.
36. Nathan Heller, "Estonia, the digital republic," *The New Yorker*, 11 December 2017, https://www.newyorker.com/magazine/2017/12/18/estonia-the-digital-republic.
37. Matt Reynolds, "'Land is so yesterday': E-residents and 'digital embassies' could replace country borders," *Wired*, 17 October 2016, https://www.wired.co.uk/article/taavi-kotka-estonian-government.

38. Elisabeth Braw, "'E-stonia' attempts to become the Uber of economies by introducing virtual residency," *Newsweek*, 30 October 2014.

17. THE EVOLUTION OF THE E-STATE

1. Viljar Lubi: Vice Minister for Economic Development 2016–2021, Ambassador of Estonia to the UK 2021–current as of publishing.
2. "Interview with Viljar Lubi, the Head of Accelerate Estonia's Board," AccelerateEstonia Blog, 28 March 2023, https://accelerateestonia.ee/interview-with-viljar-lubi-the-head-of-accelerate-estonias-advisory-board/.
3. Keegan McBride, Hendrik Lume, Gerhard Hammerschmid, and Andres Raleste, "Proactive public services—The new standard for digital governments," The Centre for Digital Governance at the Hertie School and Nortal, May 2023, https://nortal.com/wp-content/uploads/2023/06/white-paper_proactive_public_services_en.pdf.
4. Peeter Vihma, *Twenty Years of Building Digital Societies: Thinking about the Past and Future of Digital Transformation* (Tallinn: e-Governance Academy 2023), 31
5. "E-Estonia digital discussion: WTF****e, e-Estonia?," YouTube, 3 December 2021, video, 22:40, https://www.youtube.com/watch?v=6vJTutuWPv8&t=3110s.
6. "Platform for proactive government services," Proactive Government Services, Republic of Estonia Information System Authority, https://www.ria.ee/en/state-information-system/central-platforms-provision-public-services/proactive-government-services.
7. Andy Kessler, "Estonia's lessons for Ukraine," *The Wall Street Journal*, 8 October 2023, https://www.wsj.com/articles/lessons-for-ukraine-from-estonia-corruption-economy-taxes-education-business-investment-483f974b.
8. "Tackling the time tax: How the federal government is reducing burdens to accessing critical benefits and services," Office of the President of the United States, July 2023, https://www.whitehouse.gov/wp-content/uploads/2023/07/OIRA-2023-Burden-Reduction-Report.pdf.
9. "Tackling the time tax: How the federal government is reducing burdens to accessing critical benefits and services," Office of the President of the United States, July 2023, https://www.whitehouse.gov/wp-content/uploads/2023/07/OIRA-2023-Burden-Reduction-Report.pdf.
10. "Tackling the time tax: How the federal government is reducing burdens to accessing critical benefits and services," Office of the President of the United States, July 2023, https://www.whitehouse.gov/wp-content/uploads/2023/07/OIRA-2023-Burden-Reduction-Report.pdf.
11. Andero Kaha, "Report: Generative AI can boost Estonia's GDP by up to 8%…," *e-Estonia*, 25 April 2024, https://e-estonia.com/report-generative-ai-can-boost-estonias-gdp-up-to-8.
12. Marten Kaevats, "Estonia considers a 'kratt law' to legalise artificial intelligence

(AI)," E-Residency Blog, 25 September 2017, https://medium.com/e-residency-blog/estonia-starts-public-discussion-legalising-ai-166cb8e34596.
13. Brian Straight, "Company behind Walmart's Pickup Towers is back with autonomous delivery vehicle," *FreightWaves*, 15 April 2021, https://www.freightwaves.com/news/company-behind-walmarts-pickup-towers-is-back-with-autonomous-delivery-vehicle.
14. Marten Kaevats, "Estonia considers a 'kratt law' to legalise artificial intelligence (AI)," E-Residency Blog, 25 September 2017, https://medium.com/e-residency-blog/estonia-starts-public-discussion-legalising-ai-166cb8e34596.
15. "AI and the Kratt* momentum," *e-Estonia*, 11 October 2018, https://e-estonia.com/ai-and-the-kratt-momentum/.
16. "Estonia will have an artificial intelligence strategy," News, Republic of Estonia Government Office, 27 March 2018, https://www.riigikantselei.ee/en/news/estonia-will-have-artificial-intelligence-strategy.
17. "The virtual assistant Bürokratt," AI, https://www.kratid.ee/en/burokratt.
18. "Estonia's national artificial intelligence strategy 2019–2021," Government of the Republic of Estonia, July 2019, https://f98cc689-5814-47ec-86b3-db505a7c3978.filesusr.com/ugd/7df26f_27a618cb80a648c38be427194affa2f3.pdf.
19. "Estonia's national artificial intelligence strategy 2019–2021," Government of the Republic of Estonia, July 2019, https://f98cc689-5814-47ec-86b3-db505a7c3978.filesusr.com/ugd/7df26f_27a618cb80a648c38be427194affa2f3.pdf.
20. "Estonia will have an artificial intelligence strategy," News, Republic of Estonia Government Office, 27 March 2018, https://www.riigikantselei.ee/en/news/estonia-will-have-artificial-intelligence-strategy.
21. Keegan McBride, "Digital bureaucracies: How data and artificial intelligence are transforming our governments," *Apolitical*, 22 January 2024, https://apolitical.co/solution-articles/en/how-data-and-artificial-intelligence-are-transforming-our-governments.
22. "Esotnia's CIO on how the country remains a digital leader in Europe," *Silicon Republic*, 14 January 2022, https://www.siliconrepublic.com/enterprise/estonia-digital-nation-tech-siim-sikkut.
23. "Estonia AI strategy report," European Commission, https://ai-watch.ec.europa.eu/countries/estonia/estonia-ai-strategy-report_en#ai-to-address-societal-challenges.
24. "AI use cases," AI Empowering Estonia, https://www.kratid.ee/en/ai-use-cases.
25. Milda Aksamitauskas, "Reflections from a European conference on data standards," Georgetown Beeck Center, 16 November 2023, https://beeckcenter.georgetown.edu/reflections-from-a-european-conference-on-data-standards/.
26. "Vision—Bürokratt: The next level of digital state development in e-Estonia," AI Empowering Estonia, https://www.kratid.ee/en/burokratt-visioon.

27. William D. Eggers, Michele Causey, David Noone, Pankaj Kishnani, and Mahesh Kelkar, "The digital citizen: US survey of how people perceive government digital services," *Deloitte Insights*, 2023, https://www2.deloitte.com/content/dam/insights/articles/in176014_cgi_digital-citizen-survey/DI_Digital-citizen-us.pdf.
28. Indrek Mäe, "Estonia's Bürokratt, a concept of how state could operate in the age of artificial intelligence...," *Invest in Estonia*, May 2023, https://investinestonia.com/estonias-burokratt-is-a-concept-of-how-state-could-operate-in-the-age-of-artificial-intelligence/.
29. Débora Lopes Gonçalves, "Digital public services based on open source: Case study on Bürokrat," European Commission Open Source Observatory, 11 March 2022, https://joinup.ec.europa.eu/collection/open-source-observatory-osor/document/digital-public-services-based-open-source-case-study-burokratt.
30. Author interview with Ott Velsberg, Tallinn, 17 April 2024.
31. Ott Velsberg, "Estonia's AI vision: Building a data-driven society and government: An interview with Dr. Ott Veslberg, the Chief Data Officer of Estonia," *Apolitical*, 12 February 2024, https://apolitical.co/solution-articles/en/estonias-ai-vision-building-a-data-driven-society-and-government?utm_campaign=Weekly%20briefing%20%E2%80%94%20Platform&utm_medium=email&_hsmi=82885414&_hsenc=p2ANqtz--69P1EEASJ0ybji4xn-2BbQXfB_rWwPoowJPoT5YoMD5h2OfIMT-jo8x8bZRFxyuU2voNszAf1yHz8HMFI4ZD9udCZy_Q&utm_content=82873920&utm_source=hs_email.
32. Robyn Scott, "Transforming public sector services using generative AI: Global case studies," interview by Olivia Neal, *Public Sector Future*, Microsoft, https://wwps.microsoft.com/episodes/apolitical-gen-ai.
33. Robyn Scott, "What governments could learn from startups on AI," *Sifted*, 9 February 2024, https://sifted.eu/articles/government-learnings-startups-ai.
34. Ott Velsberg, "Estonia's AI vision: Building a data-driven society and government: An interview with Dr. Ott Veslberg, the Chief Data Officer of Estonia," *Apolitical*, 12 February 2024, https://apolitical.co/solution-articles/en/estonias-ai-vision-building-a-data-driven-society-and-government?utm_campaign=Weekly%20briefing%20%E2%80%94%20Platform&utm_medium=email&_hsmi=82885414&_hsenc=p2ANqtz--69P1EEASJ0ybji4xn-2BbQXfB_rWwPoowJPoT5YoMD5h2OfIMT-jo8x8bZRFxyuU2voNszAf1yHz8HMFI4ZD9udCZy_Q&utm_content=82873920&utm_source=hs_email.
35. Daniel Vaarik, "Where stuff happens first: White paper on Estonia's digital," Internet Archive, https://web.archive.org/web/20201101071644/https://www.mkm.ee/sites/default/files/digitalideology_final.pdf.
36. Joe Burton, "AI: the real threat may be the way that governments choose to use it," *The Conversation*, 2 November 2023, https://theconversation.com/ai-the-real-threat-may-be-the-way-that-governments-choose-to-use-it-216660.
37. Joe Burton, "AI: The real threat may be the way that governments choose to use it," *The Conversation*, 2 November 2023, https://theconversation.com/ai-the-real-threat-may-be-the-way-that-governments-choose-to-use-it-216660.

38. Sabrina Siddiqui and John D. McKinnon, "Biden Administration warns of AI's dangers. There's a limit to what White House can do," *The Wall Street Journal*, 4 May 2023, https://www.wsj.com/articles/white-house-warns-of-risks-as-ai-use-takes-off-d4cc217f.
39. Marc Andreessen, "The techno-optimist manifesto," Andreessen Horowitz Blog, 16 October 2023, https://a16z.com/the-techno-optimist-manifesto/.
40. "Luukas Ilves, Gulsanna Mamediieva, and David Eaves, with Vice Prime Minister of Ukraine Mykhailo Fedorov," The Rockefeller Foundation, https://www.rockefellerfoundation.org/bellagio-conversations/luukas-ilves-gulsanna-mamediieva-and-david-eaves-with-vice-prime-minister-of-ukraine-mykhailo-fedorov/.
41. Ott Kaukver, "Oral history interview with Ott Kaukver," *Baltic Video Testimonies at Stanford Libraries*, 26 August 2017, video, 32:17, https://purl.stanford.edu/hv021wq1614 (lightly edited for clarity).
42. "AccelerateEstonia," Apply, AccelerateEstonia, https://accelerateestonia.ee/apply/.
43. Dea Paraskevopoulos, "CIO of Estonia, Siim Sikkut, opens the country's digital government up as a playground," *e-Estonia*, 21 October 2020, https://e-estonia.com/cio-of-estonia-siim-sikkut-opens-the-countrys-tech-stack-to-the-world/.
44. "AI I nationens intresse—hela filmen (5 timmar), YouTube, 25 February 2024, video, 31:47, https://www.youtube.com/watch?v=8A-iGNa80Mk.
45. Scaleup: a company that is generally in between startup and mature enterprise and defined as one that has a clear product market fit and is "scaling."

CONCLUSION

1. "Kaja Kallas—Building personalized government," YouTube, 20 June 2023, video, 0:35, https://www.youtube.com/watch?v=gfEnysn__q8.
2. Kattel, Rainer and Ines Mergel, "Estonia's digital transformation: Mission mystique and the hiding hand," in Paul 't Hart and Mallory Compton (eds), *Great Policy Successes* (Oxford, 2019; online edn, Oxford Academic, 24 October 2019), https://doi.org/10.1093/oso/9780198843719.003.0008, accessed 13 May 2024.
3. Jennifer Pahlka, "The public wants I-95-ness," *Eating Policy*, 27 February 2024, https://eatingpolicy.substack.com/p/the-public-wants-i-95-ness.
4. Marc Andreessen, "Marc Andreessen on AI and dynamism," interview by Tyler Cowen. Conversations with Tyler, 13 March 2024, https://conversationswithtyler.com/episodes/marc-andreessen-2/.
5. K. Sudhir, Shyam Sunder, and Ted O'Callahan, "Faculty viewpoints: What happens when a billion identities are digitized?," *Yale Insights*, 27 March 2020, https://insights.som.yale.edu/insights/what-happens-when-billion-identities-are-digitized.
6. K. Sudhir, Shyam Sunder, and Ted O'Callahan, "Faculty viewpoints: What

happens when a billion identities are digitized?," *Yale Insights*, 27 March 2020, https://insights.som.yale.edu/insights/what-happens-when-billion-identities-are-digitized.

7. Lee Kuan Yew, *From Third World to First: The Singapore Story: 1965–2000* (New York, NY: HarperCollins Publishers 2000), 144.
8. "E-Estonia digital discussion: WTF****e, e-Estonia?," YouTube, 3 December 2021, video, 45:39, https://www.youtube.com/watch?v=6vJTutuWPv8&t=3110s.
9. Walter R. Iwaskiw (ed.), "Estonia, Latvia, and Lithuania: Country Studies," Library of Congress, January 1995, https://www.marines.mil/Portals/1/Publications/Estonia,%20Latvia,%20and%20Lithuania%20Study_1.pdf.
10. Don Moynihan and Gulsanna Mamediieva, "Digital resilience in Ukraine," *Can We Still Govern?*, 31 October 2023, https://donmoynihan.substack.com/p/digital-resilience-in-ukraine.
11. Based on off-the-record conversations.
12. Ann-Marii Nergi, "World-class IT R&D in Estonia: Cybernetica creates IT-solutions that just cannot fail," Life in Estonia, November 2020, https://investinestonia.com/world-class-it-rd-in-estonia-cybernetica-creates-it-solutions-that-just-cannot-fail/.
13. Marten Kaevats, "How to build digital public infrastructure: 7 lessons from Estonia," *World Economic Forum*, 4 October 2021, https://www.weforum.org/agenda/2021/10/how-to-build-digital-public-infrastructure-estonia/ (lightly edited to correct spelling error).
14. Sten Tamkivi, "Lessons from the world's most tech-savvy government," *The Atlantic*, 24 January 2014, https://www.theatlantic.com/international/archive/2014/01/lessons-from-the-worlds-most-tech-savvy-government/283341/.
15. President Toomas Hendrik Ilves, "Toomas Ilves: Lessons in digital democracy from Estonia," interview by Russ Altman, *The Future of Everything*, Stanford University School of Engineering, 4 February 2019, video, 9:09, https://www.youtube.com/watch?v=ISchpVmmQiU.
16. Aneesh Chopra, *Innovative State: How New Technologies Can Transform Government* (New York: Atlantic Monthly Press, 2016), 25.
17. Aneesh Chopra, *Innovative State: How New Technologies Can Transform Government* (New York: Atlantic Monthly Press, 2016), 45.
18. Mike Bracken, "No more 'Big IT': The failed 90s model has ruined too many lives," *Financial Times*, 19 January 2024, https://www.ft.com/content/6b1f9cf2-8ab7-483e-8570-febc08f8f1e3.

STEN TAMKIVI, INVESTOR AND ENTREPRENEUR, ON SKYPE, ESTONIA'S UNIQUE STARTUP CULTURE, AND TECH ECOSYSTEM DEVELOPMENT

1. In-person interview with Sten Tamkivi in Tallinn on 9 April 2024.

EXPLORING THE UNIQUE LINKS BETWEEN ESTONIA AND FINLAND

1. Interview with Matt Mitchell on 4 April 2024.
2. Neil Taylor, *Estonia: A Modern History* (London: C. Hurst & Co. 2018), Kindle edition, 176.

GETTING THINGS DONE IN GOVERNMENT WITH MARINA NITZE

1. Interview with Marina Nitze on 19 March 2024.

HIGHLIGHTS FROM A DISCUSSION WITH FORMER U.S. DEPUTY CTO JENNIFER PAHLKA AND AUTHOR OF *RECODING AMERICA* ON STATE CAPACITY

1. In-person interview in Washington, D.C., with Jennifer Pahlka on 15 March 2024.

HOW THE NEXT GENERATION OF POLICYMAKERS VIEW THE FUTURE

1. In-person interview with MP Hanah Lahe in Tallinn on 9 April 2024.
2. Shafi Musaddique, "'People put a lot of hope on me': Estonia's youngest MP already making waves," *The Guardian*, 28 February 2024, https://www.theguardian.com/world/2024/feb/28/estonia-youngest-mp-hanah-lahe-interview.

A DISCUSSION WITH SIIM SIKKUT, FOUNDER OF DIGITAL NATION AND FORMER CIO OF ESTONIA ON THE IMPACT AND POTENTIAL OF AI IN GOVERNMENT

1. In-person interview with Siim Sikkut in Tallinn on 22 April 2024.
2. "Kaja Kallas: E-governance is a question of building up democracy," *The Baltic Times*, 30 May 2023, https://www.baltictimes.com/kaja_kallas__e-governance_is_a_question_of_building_up_democracy/.

INDEX

Note: Page numbers followed by "*n*" refer to notes.

Aadhaar platform, 222
Aaviksoo, Jaak, 110–11, 112, 146, 147, 233
Accelerate Estonia (program), 123, 215–16
Acton, Brian, 170
Adam Smith Institute, 52
Africa, 1, 141
AI (Artificial Intelligence), 4–5, 71, 94, 154, 205, 227
   impacts and potential of, 7, 166, 212, 255–8
   rise of, 204
   risks to humanity, 24
   services and products, 209–14
AI virtual assistants, 256
AI-enabled drones, 220
Air Force Research Laboratory, 132
Allandi, Marge, 31
Altman, Sam, 171
Amazon Web Services (AWS), 149, 199–200
Amazon, 6, 172, 200, 223
American Minutemen, 147
Americans, 93, 115, 171
Andreessen Horowitz, 107, 118, 184
Andreessen, Marc, 106, 107, 153, 214, 221–2
Anduril, 65
Annus, Ruth, 187–8
Annus, Toivo, 14, 17–18
Ansper, Arne, 95, 120
Apple Music, 15
Ars Technica, 17, 20–1
Article (V), 148
Atlantic (magazine), 83, 101
Atlantic Council, 97, 147
Atomico (investment firm), 118, 119, 126
Avoid Legacy Technology (principle), 85, 87–8
Azerbaijan, 194, 196

Baker, James, 39
Baltic Sea, 13
Baltic states, 40, 190, 234
Baltic Way (Human Chain), 34–6
Barbados Welcome Stamp initiative, 172
Barbados, 172

321

INDEX

Benin, 132, 222
Berlin Wall, 11, 58
Bertelsmann survey (2019), 104
Big Con: How the Consulting Industry Weakens Our Businesses, Infantilizes Our Governments and Warps Our Economies, The (Collington and Mazzucato), 156
Bina, Eric, 107
birth certificate, 83–4
Blackwater, 175
Blank, Steve, 167
Bloomberg, 28, 78, 174
Bloomberg, Mike, 97
Bluemoon Interactive, 13–14, 15
Bolt, 1, 12, 25, 111, 118, 126
Bracken, Mike, 227
Britain, 81, 156
Buell, Duncan, 96
Build vs. Buy and Embrace Frugality (principle), 85, 87–8
bureaucracy, 54–5, 58–9, 137–8, 192, 249
    hacking, 243–5
    impacts on US, 167–8, 181, 209
    Once-Only principle, 86–7
    risks of, 61–2
Burn Book (Swisher), 18
Bürokratt, 211–12, 255–6
Burton, Joe, 214
Buterin, Vitalik, 203

Canada, 70, 104
CCDCOE. *See* Cooperative Cyber Defence Centre of Excellence (CCDCOE)
Centers for Disease Control and Prevention, 167
Centre for the Study of Existential Risk, 24
CERT. *See* Computer Emergency Response Team (CERT)

ChatGPT, 4, 25, 170–1, 209, 213
Chechens, 224
Chernobyl nuclear disaster (1986), 29
China, 6–7, 220
Chinese Communist Party, 220
Chopra, Aneesh, 226–7
Christmas, 28
Cleveron (701), 210
climate change, 6, 248, 252
Clinton, Bill, 66
CloudZero, 152
CNBC (news channel), 171
CNET, 21
Coinbase, 20
Collington, Rosie, 156
Colombia, 132, 222
Computer Emergency Response Team (CERT), 146
computing, 68, 71, 106, 166, 180
Cooperative Cyber Defence Centre of Excellence (CCDCOE), 148–51
Corruption Perceptions Index, 139
COVID-19 pandemic, 104–5, 171, 173
Cowen, Tyler, 222
Crimea, 41
Crimean Black Sea, 40
cryptocurrency, 12, 195
cryptography, 130
"culture eats policy", 249
Cyber Defence Unit, 147
cyberattack, 83, 90–1, 143–4, 168
    impacts of, 146–51, 152–5, 201–2
    Russian cyberattacks on Ukraine, 199–200

INDEX

Cybernetica, 95, 129–33, 134, 140–1
cybersecurity, 91, 96
   investments in, 151–2, 157, 204
cyberspace, 90
Cyprus, 171

Dallas (TV show), 42
Davidson, James Dale, 169
Delhi, 192
Denmark, 13
Department of Homeland Security, 209
DFJ, 20–1
DIANA (Defence Innovation Accelerator for the North Atlantic), 123, 151
digital age, 8, 72, 180, 183, 220, 302n12
Digital by Default (principle), 85, 87–8
digital era, 94, 109, 110, 116, 159, 165
Digital Government Excellence (Sikkut), 8, 137
digital identity (e-ID), 77, 79–82, 83–4, 88, 143, 188
   banking sector, 91–2
   creation of, 222, 224–5
   e-resident digital identity cards, 190–1
   medical emergency, 101–3
digital infrastructure, 91–2, 122, 145, 152, 189–90, 199, 222
   access to the youth, 116
   failure in creation of, 182
digital public infrastructure, 154, 156, 299–300n2
Digital Signatures Act, 79, 80, 88
digital signature, 80, 83, 94, 95, 131, 191

digital skills, 109, 112–13, 223
digital world, 113, 155, 168, 177, 183
   mobile devices as primary interface to, 71
   preparation of youths to, 122
   risks in, 154, 199
Diia (application), 198–9, 201, 217
Docusign, 80
Draper International, 20
Draper, Bill, 19–20
Draper, Tim, 20, 180, 184, 190–1, 192
Draper, William Henry, 19
Drechsler, Wolfgang, 106
Drucker, Peter, 159
Dutch Supreme Court, 16
Dynasty (TV show), 42
Dzene, Inita, 35

e-ambulance services, 101
East Asia, 203
Eastern Europe, 34, 64
Eaves, David, 182
eBay, 21, 23, 118, 170
e-commerce, 6
Economist, The (newspaper), 191, 201
Eesti Televisioon, 29
e-Estonia Briefing Center, 165
e-Governance Academy, 8, 57, 69, 112–13, 150, 155, 198
e-government services, 57, 68, 77, 91, 129, 154, 168
   online access in Ukraine, 197
   use of, 208
E-Government Success Stories: Learning from Denmark and Estonia", 131–2
e-government systems, 213, 215, 222, 225

INDEX

development of, 7, 78, 85, 121, 132, 158, 206
   role in minimizing corruption, 69, 139
e-health system, 89, 92, 99–102, 135
Ehin, Piret, 98
EHR. *See* electronic health records (EHR)
EKRE (political party), 96–7, 253
electronic health records (EHR), 100–1
environmental issues, 251–2
e-prescription program, 99
e-prescriptions, 101
e-Residency program, 183, 187–8, 195–6, 222–3, 225, 277n10
   development and growth, 189–94
"Estonia, the Digital Republic", 77
Estonia: A Modern History (Taylor), 56
Estonian Air Navigation Services, 131
Estonian Defence League, 147
Estonian Development Fund, 187
Estonian Parliament, 252
Estonian Tax and Customs Board, 93
Estonian World (online magazine), 8, 105
Estonians' Liberation Way (Made), 37
Ethereum, 203
Europe, 1, 6, 7, 12, 22–3, 133, 197
European Central Bank, 195
European Commission, 104, 139, 158
European Parliament elections (2019), 251–2
European Parliament, 252
European Union, 12, 57, 69, 99, 139, 192, 220
eVorog, 201

Facebook (Meta), 136, 170
FAFSA, 248–9
Farivar, Cyrus, 146, 190
FastTrack, 16
Fedorov, Mykhailo, 198, 200, 214–15
Fijian island, 174
Financial Times (newspaper), 172, 179, 227
Finland, 92, 108, 234
   introduced digital identity system, 79
   relationship with Estonia, 1, 13, 32, 57, 65, 99, 121, 159
Florida, 241
Forbes (magazine), 175, 192
France, 139, 182
Free to Choose (Milton and Rose Friedman), 51
Freedom of Information Act, 81
Freedom Square, 12
Friedman, Milton, 51–2, 208
Friis, Januus, 14–16
From Third World to First (Lee Kuan Yew), 61
Future Investment Initiative, 174–5
Future of Life Institute, 24

Gates, Bill, 4, 66
GDP, 52, 55, 59, 92, 119, 210, 220
geopolitical changes, 7, 169, 184
Georgia, 148, 172
German Marshall Fund, 148
Germans, 241
Germany, 82, 107–8, 126, 192

INDEX

adaption to industrial age, 166, 168, 179, 182
CCDCOE formed by, 149
digital age failure, 220
tax code, 92–3
GI Bill, 244
GitHub, 171
GitLab, 213
glasnost (openness), 28, 38, 40
Global Cybersecurity Index (2022), 151
Goble, Paul, 39
Godbout, Greg, 158
Google, 21, 119, 210
Gorbachev, Mikhail, 27–8, 29, 37–40
Gorohhov, Janer, 23
Guardian, The (newspaper), 38, 66–7, 131, 145
Guardtime, 132, 133, 136, 140–1
Gulf of Finland, 241

Haataja, Samuli, 146
Hack Your Bureaucracy (Nitze and Sinai), 167
Halliburton, 175
Hansapank (Swedbank), 113
Hanseatic League, 47
Harris, Kamala, 214
Hartenbaum, Howard, 20
Harvard Business Review (magazine), 176
Harvard's Kennedy School, 87
Health Services Organization Act (2001), 103
Heinla, Ahti, 14
Heller, Nathan, 138, 196, 203
Hellrand, Maris, 31
Helsinki, 17, 42, 119
Heritage Foundation, 52
Hindriks, Karoli, 173

Hinrikus, Taavet, 22–3, 118, 123
Hirve Park (Tallin), 32
Honduras, 184
Horowitz, Ben, 106, 193
Hotmail, 20
Houthis, 168
Hryvnia (Ukrainian currency), 70
Huffington Post, 176
Hugging Face, 65
human rights, 251–2

Iceland, 44, 70, 87, 149
Identity Documents Act, 79
identity verification, 91
Ilves, Luukas, 136, 151, 216
Ilves, Toomas Hendrik, 62, 66, 108–9, 114, 127, 132
  created policies for e-state, 90
  thoughts on IT sector, 68–9
  thoughts on paper world, 154–5
  views on digital ID, 81
  See also Tamkivi, Sten
IMF, 49–50, 54, 172
Index of Economic Freedom, 60
Index Ventures, 122
India, 20, 192, 222
Indonesia, 171
information age, 166, 168–70, 175, 177, 204, 302n12
information technology, 78, 99, 151, 158
Innovative State (Chopra), 227
insular culture, 67–8
Intel, 115
Internal Security Service, 144–5
international press, 77, 104
International Republican Institute, 52
Internet of Elsewhere, The (Farivar), 146

INDEX

Ireland, 149, 234
Iron Curtain, 13, 41–2, 66, 108, 116
IRS (software), 93, 94
Israel, 223
IT industries, 109
Italy, 149
Ivanilov, Yuri, 41
i-voting (e-voting), 92, 94–9, 131, 225

Jamestown Foundation, 39
Japan, 149, 165–6, 168, 220
Jobs, Steve, 18
Joost, Jüri, 42–3
Journal of Baltic Studies, 55
Judo, 63–4
Judo: History, Theory, Practice (Putin), 63–4
Jurvetson, Steve, 20–1, 53, 69, 169, 190

Kaevats, Marten, 210, 224, 226, 255–6
Kaidro, Raul, 81
Kaljulaid administration, 210
Kaljulaid, Kersti, 55, 64, 124, 207, 224
Kallas, Kaja, 28, 54, 64, 78, 136–7, 208
  views on e-governance, 219
Kallas, Siim, 30, 50
Kalvet, Tarmo, 67, 79
Kasearu, Kaire, 157
Kasesalu, Priit, 14
Kasevāli, Uno, 42–3
Kask, Laura, 79, 202
Kattel, Rainer, 8, 52, 68, 140, 220
Kaukver, Ott, 22, 70–1, 116, 215, 234
Kawasaki, Guy, 193

Kazaa, 16–17
Kert, Johannes, 150–1
Kessler, Andy, 208
Khanna, Ro, 177
Kiribati, 174
Kissinger, Henry, 108
Koivisto, Mauno, 56–7
Kokk, Jaanus, 42–3
Kopli, Merit, 144
Korjus, Kaspar, 191, 195
Kosmonaut (computer game), 14
Kotka, Taavi, 135, 137, 179–80, 187–8
  Country-as-a-Service concept, 183–5
  roundtable discussion with Kaljulaid and Kaevats, 224
Koum, Jan, 170
Kratt, 210–11
Kroon, 49–51, 54, 60
Kukoff, Zak, 171
Kyrgyzstan, 132

Läänemets, Lauri, 158
Laar, Mart, 30, 34, 49–52, 54
  created policies for e-state, 90
  economic shock therapy, 224
  Thatcher and, 56
  thoughts on e-Residency, 194–5
  thoughts on IT sector, 68–9
  trade-not-aid, 58–9
Lagos, 180
Lahe, Hanah, 251–3
Lapõnin, Aare, 158
Latvia, 34, 49, 64, 115, 149
Lauristin, Marju, 29
Law, Innovation and Technology (journal), 146
League of Nations, 240
Lee Kuan Yew, 61, 108, 189, 224

INDEX

Lennart Meri Tallinn Airport, 1, 59
Less State is More (principle), 85
Lewin, Amy, 119
Lieven, Anatol, 30, 50
Lift99, 117
Lithuania, 34, 38–9, 42–3, 64–5, 115, 149, 194
Lithuanians, 38, 234
Locked Shields (cyber defense exercise), 150
London, 123, 184, 191, 199
Look@World campaign, 109, 112–13, 114
Los Angeles, 16
Lubi, Viljar, 206, 215
Lucas, Edward, 190–1
Luxembourg, 202–3, 236

Made, Tiit, 29, 30, 37
Mägi, Enel, 113
Maigre, Merle, 150
Malaysia, 132
Mamediieva, Gulsanna, 224–5
Marin, Sanna, 124
Mattiisen, Alo, 27
Maxwell, Liam, 200, 201, 203
Mazzucato, Mariana, 156
McBride, Keegan, 104–5, 206
McDonald's, 11
McKinsey report (2012), 89
McKinsey, 176
Mergel, Ines, 68, 140, 220
Meri, Lennart, 58–9, 64, 66, 115, 121, 228
　announced Tiger Leap program, 111–12
　death of, 127
Mexico, 47, 171
Miami, 223
Microsoft, 23, 69, 118, 149
Middle East, 141, 175

Milli, Peeter, 42–4
Millimallikas, 19
Ministry of Digital Transformation, 198–200
Ministry of Social Affairs, 99
Model T, 154
modern era, 5–6, 45, 58–9, 181
Molotov-Ribbentrop Pact, 32, 34–5
Monaco, 203
Moore, Alice, 156
Mosaic browser, 106, 107–8, 153
Moscow Olympic Games (1980), 31
Moscow, 27, 33, 35, 38, 40, 41–4
Moynihan, Don, 224–5
Mumbai, 223
Munich, 223
Musk, Elon, 4, 24–5
Mussolini, Benito, 221
My Chemical Romance (band), 2

Nagorno Karabakh, 175
Napster, 15–17
Narva, 2
National Cemetery Administration, 244
National Geographic (television network), 58
NATO, 12, 69, 110, 134, 143, 188
　response to cyberattacks, 148–51
natural resources, 63, 71–2, 108, 129, 168, 223
Naughton, John, 156
Nazi Germany, 144
Nazis, 144, 240–1
Netherlands, 184, 236
New Union Treaty, 39–40
New York Times, The (newspaper), 38, 134

INDEX

New Yorker, The (magazine), 77, 82–3, 138, 203
New Zealand, 171
Nigeria, 180–1
Nitze, Marina, 8, 167
No Land Is Alone (song), 27
Nokia, 121
Nord VPN, 235
Nordic Institute for Interoperability Solutions, 83
Nordics, 65
Nortal, 132, 133, 140–1
North America, 141
Northern Europe, 47
Not Optional (advocacy group), 122
Nothing Is True and Everything Is Possible (Pomerantsev), 148
NotPetya, 149–50

Obama administration, 57
Obama, Barack, 105, 151
Obamacare, 89
Once-Only (principle), 85, 86–7
online banking, 67, 90, 91–2, 109, 143, 152, 154
Open Estonia Foundation, 112, 113
Open Society Foundation, 57
Open Society Institute, 69–70
open-source software, 86

Pahlka, Jennifer, 57, 166, 168, 221
Palantir, 117
Parek, Lagle, 48, 271n10
Pärt, Arvo, 31
PayPal, 117, 119
peer-to-peer (P2P), 15, 17, 25
perestroika (restructuring), 28, 38, 40
Personal Data Protection Act, 79, 80–1, 88

phishing, 143
phosphorite mining, 27, 29
Piirimäe, Kaarel, 35
Pipedrive, 126
Politico, 119, 134, 151
Politics of Uncertainty: The United States, the Baltic Question, and the Collapse of the Soviet Union (Bergmane), 43
Post Office Horizon scandal, 156, 227
Postimees (newspaper), 144
Pravda (newspaper), 29
Priisalu, Jaan, 152
"Principles of Estonian Information Policy", 78, 92
privacy rights, 80
private data, 81
Prospera, 184
Prozorro, 70
Public Information Act, 79, 80–2
Putin, Vladimir, 63–4, 145

Quartz, 173, 191
Quincy Institute for Responsible Statecraft, 134

Radio Free Europe, 66
Rahva Hääl (newspaper), 109
RAND Corporation, 169
Raun, Toivo, 55, 59
Reagan, Ronald, 20
Recoding America: Why Government Is Failing in the Digital Age and How We Can Do Better (Pahlka), 57
Red Army soldiers, 144
Rees-Mogg, William, 169
Reform Party (political party), 252
Remnick, David, 40
Reuters, 40–1
Rick, Raul, 122

INDEX

Riga, 49
Riisalo, Tiit, 120–1
Rockefeller Foundation, 214–15
Rodeo: Taming a Wild Country (documentary), 54
Romer, Paul, 222
Rounds, Mike, 4
Ruble, 49–51
Russia, 39, 49, 125, 143–4, 145, 174, 202
   animosity between Estonia and, 96
   cybercrimes, 90, 146–8, 199–200
   recruitment of foreign soldiers, 175
   trade with Estonia, 51
   See also Ukraine
Russian troops, 39, 50
Russians, 27, 41, 201

Sahel, 175
same-sex marriage, 124
Sarkissian, Armen, 53–4, 168–9
Saudi Arabia, 174–5
Savisaar, Edgar, 30, 34
Schengen Information System, 158
Schmidt, Eric, 4, 167, 174–5
Science Tiger, 116
Scott, Robyn, 213
Sherman, Brad, 94
Shirreff, Richard, 202
Siberia, 28, 37
Sikkut, Riina, 102
Sikkut, Siim, 8, 125, 187–8, 193–4
   thoughts on AI, 211–12
   views on digital bureaucracy, 206
Sild, Ants, 110, 112
Silicon Valley, 5, 23, 119–20, 123, 174, 204

Sinai, Nick, 167
Singapore, 61, 108, 223–4
Singing Revolution, The, 32, 37, 41–2
Skeleton Technologies, 120, 129
Skype, 13, 69, 107, 111, 118
   rise and impacts of, 18–25
   success of, 125, 170
Slack, 171, 235
Slovak Republic, 149
Small States Club: How Small Smart Powers Can Save the World, The (Sarkissian), 53
SmartCap, 123
Snowballs (data storage devices), 199–200
social media, 211, 220
social security number, 83–4
"Software is eating the world", 153
South America, 141
South Carolina, 241
Southern Europe, 223
Sovereign Individual: Mastering the Transition to the Information Age, The (Davidson and Rees-Mogg), 169–70, 183, 204
Soviet era, 2, 11–12
Soviet troops, 37, 38, 43–4
Soviet Union, 33–5, 47–8, 60, 98–9, 144
   collapse of, 11, 14
   coup (1991), 37
   New Union Treaty, 39–40
   phosphorite mining activities, 27–9
Soviets, 13, 28, 31, 42, 63, 241
Spain, 139, 149, 192
SplitKey technology, 131
Spotify, 15
Srinivasan, Balaji, 184, 203
St. Petersburg, 44
Stanford Library, 8, 139

INDEX

Starlink, 181
Starship Robotics, 120, 210
Startup Estonia, 123–4, 126
"State in the Smartphone" concept, 217
Stavridis, James, 175
Stockholm, 17
Stoltenberg, Jens, 148
stone age, 152, 154
Stripe Atlas, 180, 182, 195
surveillance cameras, 5
Sutter Hill Ventures, 19–20
Sweden, 13, 17, 52, 107–8, 253
Swedes, 240

Tallinn Old Town Days (festival), 32
Tallinn Town Hall square, 113
Tallinn TV Tower, 42–4
Tallinn, 11–12, 29, 95, 174, 241–2
 arrest of the putschists in, 43
 creation and accreditation of CCDCOE, 148
 e-Governance Academy in, 57, 69–70
 Human Chain, 34–5
 protests in, 32
 Skype operations in, 21
 spike in crimes rate, 47–8
Tallinn, Jaan, 13–14, 15, 23–5, 66
Tamkivi, Sten, 23, 83, 103, 110, 117–18, 226
 advisory role to Ilves, 135–6
 e-Residency program, 187–8, 193
 thoughts on Estonia's growth, 69
Tammer, Tonu, 96
Tammet, Tanel, 110
Tartu, 29, 95, 193
tax administration, 54, 92

Tax Competitiveness Index, 60
tax reviews, 94
"Teach to the Future" training program, 115
tech mafia, 117–19, 126
TechCongress, 4
Technology Neutrality (principle), 85
Tehnopol, 123
Tele2, 14–15
telecom, 109
Teles, Steven, 166
Tesla, 20
Texas, 170
Thatcher, Margaret, 56
Thiel, Peter, 169, 184, 203
Tiger Leap (campaign), 57, 109–12, 113–16, 122, 146
Tiger Robotics, 116
Timbro, 52
Time (magazine), 199
Titma, Miik, 30
TOM (*Täna Otsustan Mina*), 97–8
Topol, Eric, 100–1
Toronto, 174
TransferWise (Wise), 22, 111, 118, 126
Trembita, 150
Trump, Donald, 247
Tulsa Remote program, 172
Tunisia, 132
TurboTax, 93
Tusk, Bradley, 97
Twenty Years of Building Digital Societies (Vihma), 206
Twitter, 124
2034: A Novel of the Next World War (Ackerman and Stavridis), 175

U.S. Chamber of Commerce, 176
U.S. Cyber Command, 149
U.S. Digital Service, 167
U.S. Library of Congress, 56

INDEX

U.S. Social Security Administration, 176
Ubar, Raimund, 68
Uber, 1, 118, 223
Uganda, 82
Ukraine, 132, 136–7, 175, 195, 198–9, 224–5
   digital leadership, 216–17
   digital sovereignty, 200–04
   investment in e-services, 194
   joined with CCDCOE, 149
   launched open-source Prozorro procurement platform, 7
   Russian invasions of, 151, 184, 197, 202
   war in, 5, 125, 133, 201
Ukrainian troops, 197
Ukrainians, 150, 197, 199, 201
United Nations Conference on Trade and Development, 192
United Nations Development Programme, 69–70
United Nations, 3
University of Cambridge, 24
University of Tartu, 35
US (United States), 5, 16, 132, 139, 241
   adaption to information age, 166, 168
   digital age failure, 182, 220
   government IT projects, failure of, 157–8
   Meri visits to, 66
   tax code, 92–3
USSR State Council, 44

Väärtnõu, Oliver, 225–6
Vahi, Tiit, 50, 92, 271n17
Vahtras, Liina, 192
Vainik, Mikk, 215
Valk, Heinz, 30, 32, 33, 36, 41–2

Vatter, Ott, 191
Velliste, Trivimi, 30, 33
Velsberg, Ott, 157, 212–13
Veriff, 23
Vesilind, Priit, 58
Victory Day (holiday), 144
Viik, Linnar, 53, 66–7, 77, 79, 108, 131
Viirg, Andrus, 189
Villig, Markus, 12, 119, 125
Vilnius TV Tower, 42
Vilnius, 35, 38, 42
Vinted, 235
Viru Gates, 11
Virumaa (region), 27
Voice of America (radio network), 66
Vote With Your Phone: Why Mobile Voting Is Our Final Shot at Saving Democracy (Tusk), 97
Vox (news site), 93, 169

Wagner Group, 175
Wall Street Journal (newspaper), 208
War in the Woods: Estonia's Struggle for Survival (Laar), 51
War of Independence Victory Column, 12
War With Russia—An Urgent Warning from Senior Military Command (Shirreff), 202
Washington Post, The (newspaper), 34, 38, 84
Washington, 43, 135
"Web War One", 143, 155
Western Europe, 100, 202, 234, 311n52
Western world, 6–7, 89, 150, 220–1, 222, 227
WhatsApp, 13, 170, 209–10
White House, 43, 209, 249

INDEX

Wolfe, Josh, 65
World Bank, 49
World Cybercrime Index, 90
World Health Organization, 104
World War II, 2, 11, 13, 60, 144, 201
World Wide Web, 66
Wozniak, Steve, 24–5

X-Road, 77, 82–3, 84, 88, 92, 103, 130

Yang, Andrew, 24–5
Yeltsin, Boris, 39, 41, 43
Yemen, 168
youth issues, 251–2
Yudkowsky, Eliezer, 24

Zelensky, Volodymyr, 70, 150, 197–8, 225
Zennström, Niklas, 14–17, 19, 21, 126
Zoom, 13, 171